MATH WITH NO MATH

Unconventional Wisdom for the Capable, Perplexed,
Incredulous, Curious, and Earnest High School Math Student

MATH WITH NO MATH

Unconventional Wisdom for the Capable, Perplexed, Incredulous, Curious, and Earnest High School Math Student

Joe Poirier

PEQUOSSETTE
PUBLISHING

Math with No Math
Unconventional Wisdom for the Capable, Perplexed, Incredulous, Curious, and Earnest High School Math Student

Author and Illustrator
Written by: Joseph J. Poirier
Illustrations by: William M. Bradbury
Figures by: Joseph J. Poirier

Identifiers
Library of Congress Control Number: 2022917487
Paperback ISBN: 979-8-9866539-0-7
eBook ISBN: 979-8-9866539-1-4

PEQUOSSETTE
PUBLISHING

Contents

Acknowledgments

Inspiration for this book came from many high school math students and other people over the years—certainly too many to mention here. However, I would like to thank the following people for their painstaking review of my draft manuscript and valuable feedback while they were juggling work, family, and/or school obligations at the time: Cindy Atoji Keene, Matt Nugent, Tim Roche, Wilson Vo, Dennis Stringer, and Patty Mahoney. Artist Bill Bradbury was very understanding and flexible during the illustration process. Robin Wrighton and artist Diana Zipeto also contributed helpful information and suggestions. The responsibility for content and any errors or omissions is all mine.

1

Introduction

'It is not possible' you write me? That is not French.
- Napoleon Bonaparte, French general and emperor, 1769-1821

I hate math.

I do OK in math, but I don't see its relevance.

I do well in all my other subjects except math.

I don't get math; not that I can't do it or understand it, but what it is from a 'big-picture' perspective, and why they keep pushing it on us.

If these sound like you, in whole or in part, then this book represents my attempt at addressing your concerns and answering your questions. I've heard these phrases many times from my years as a volunteer tutor, in materials I've read, and even what I've said myself. This thinking is a problem for American society at large[1]. It is not new, but its potential negative impact is more profound than in past years given:

- The unrelenting rise in sophisticated technology, and the need to understand and manage it

---

[1] This book is directed at an American audience. However, if you are a not an American student, I think you can easily make the adaptation to your country and situation.

- Plenty of unfilled jobs requiring basic-to-advanced math skills
- Reduced employment options for those without basic-to-advanced math skills
- Too many students taking remedial math courses when they could be progressing in other ways
- Students not knowing what math is, what to do with it, why it's important, how to study and learn it properly, yet having more than enough intellectual capacity

This book is not for all students or the general public. It is, however, directed at the capable but incredulous or confused high school student. There's a bonus for the target audience so as to not detract from the message: Other than two minor exceptions where I mention the number π and provide example illustrations of condensed notes as a study technique, no math is used in this book!

Purpose

The main purpose of this book is to influence you, not to make you love math (or mathematics; the terms are used interchangeably throughout the book), not to hate it, but to understand and appreciate it in a way that most teachers, parents, guardians, and authors don't convey. I will explain what math is, what to do with it, the reason it is emphasized and *imposed* on you, give you reasons to persevere, and how to learn it properly, all in the hope that you will understand that it is not wasteful or impossible to master. I will also pass on wisdom by telling you what I wished someone had told me when I was a high school student, to provide a different perspective, and give you some direction and sense of purpose when it comes to math. What differentiates this book from others that target you regarding math is that I will explain and preach with a certain degree of nontraditional authority, and do so in an alternative, down-to-earth, simple way for young students like you to understand and, hopefully, appreciate.

This book is not just about math but also, more subtly, about a balanced education. An education without an adequate level of math is unbalanced. Math is usually lacking for the wrong reasons. All other parts of a formal education along with math, including science, arts, music, humanities, and physical education, are necessary to navigate, survive, and thrive in our economic, political, and physical world.

Some math teachers (certainly not all) don't always explain concepts completely, address fundamental questions, or make themselves available. They may never have struggled in the same way as you. They were probably always "good at math", which is one reason they're teaching it. They may have trouble empathizing with you; I will try to do so.

Failure certainly has its place in the learning process and in attaining wisdom, but some failures are excusable and inevitable, whereas others are not. Your job should be to watch others do the wrong thing and make mistakes, then learn from theirs to minimize yours. Don't text and drive. Don't stick your key in the light socket. Don't spit in the wind. And don't think wrongly about math.

Before digging deeper, let's take a step back, define you, then I'll give a background of myself to provide some credibility as an author.

Who You Are Not

You do not love math. You are not a math whiz who always and effortlessly gets *A*s in the subject. Stress-free, perpetually content, and impulsive are not adjectives that describe you. You do not thoughtlessly accept math as it is normally taught and presented. You don't operate without questioning.

Who You Are

You are a capable, curious, and mindful high school student who gets good grades in every subject except, perhaps, in math. Math sometimes makes you disillusioned or incredulous. You're open to ideas and suggestions and are not averse to schoolwork and homework. The reason you are reading this book is because you are not giving up on math. You want to improve and understand the subject more deeply, and want to see math from a different viewpoint. Your long-term goals are somewhat foggy, which is normal for a high school student. You may or may not be college bound. You are potentially a future scientist, financial professional, teacher, tradesperson, engineer, farmer, mother, or father, but certainly a taxpayer, voter, and consumer of goods and services; that is, a person who will make critical judgments with profound future implications, personally or otherwise. Lastly, you're an emotional and thoughtful human being who wants to know the reasons for doing something before being directed to do so. Which brings us to who I am not and who I am.

Who I Am Not

I am not a teacher with teaching credentials, at least not yet. In addition, I'm not the smartest guy in town, world's greatest engineer, Nobel prize winner, a nerd overly focused on math, or a widely published author (so far anyway).

Who I Am—*My Mathematical Journey*

You should want to know who the author is of any nonfiction book and if he or she has a reasonable degree of experience and authenticity. Therefore, the intent of the next sections is not to be a self-centered, long-winded biography, but to give this book some standing. I think the next sections will be highly relevant to you. You may see yourself here and there, and they will show how my way of thinking has evolved, especially about math. These sections also give you an opportunity to learn from my mistakes, insights, strengths, weaknesses, and experiences.

General. I'm a 61-year-old math advocate, who also happens to be an ex-Marine and mechanical engineer with a pretty good memory, with lots of informal experience tutoring middle and high school students. Believe it or not, I was a teenager at one time. I remember junior high school (now called middle school) and high school well, especially what I was thinking and feeling at the time. I continue to do some informal tutoring, which helps me stay engaged with young students and their math anxieties. Many aspects of youth culture have changed over the years and many haven't. I'm a generalist who is *usually* very good at empathizing with others, young or old, and find everything interesting and potentially useful. Like many, I find focusing on a particular subject to be difficult at times, in a general sense, not because of attention deficit disorder, but because I want to know and do everything. I have multiple projects going on at any one time both at work and home, since I tend to be a curious multitasker. I'm an inveterate reader, am never bored, have plenty of personal regrets but no general career-oriented ones.

Growing Up in the 1960s & 1970s. I was a very active and athletic kid and tried all sports. I excelled at none of them but could hold my own. *Jack of all trades and master of none* described me well and still does. I loved school, learning, and libraries, and read innumerable books on my own volition for learning enjoyment. I grew up with four sisters and no brothers but compensated for the no-brother situation with a lot of friends, which was easy in those days since we all came to be at the tail end of the post-World War II *Baby Boom*. I had mostly good friends. There were no personal computers or cell phones to distract us then, so we actually went outside for unstructured and unsupervised play. We had a blast.

I don't come from an academic family. We were middle class—not wealthy, not poor. I was very fortunate to have had good parents. They were high school graduates. My father was always employed full time (and periodically took on additional part-time work), and my mother stayed at home until we grew older, transitioning from part-time work to full-time by my high school years. Unfortunately, my parents were not always very instructive, patient, or academically empathetic. They harped on us kids to do our homework, but didn't follow up or monitor since many times they were distracted and hypnotized by television (TV) in the evenings.

A professional career requiring a college degree was not a big driver for me. But deep down, I wanted to be an astronaut, scientist, engineer, and/or military officer, among other things. These were the professions I respected. Joining the military was a given; there was no debate about that. I grew up in the age of NASA's[2] Mercury, Gemini, and Apollo programs. The Vietnam War was always in the news, but it was mostly an abstraction to me while I was in elementary school. I witnessed the first lunar landing, accomplished by, in my mind, virtuous, intelligent, brave, capable, manly, military men. They were technical, analytical problem solvers, who needed to have extensive knowledge plus mental and physical endurance and self-discipline to do their jobs. The astronauts could deal with technical complexity, uncertainty, prolonged physical discomfort, and chaos, not be intimidated by them, and remain calm under pressure.

Junior High School, 1973-1976. Junior high school started out great academically, but my grades progressively declined from the seventh grade to ninth. Grades of *A*s and *B*s slowly transitioned into *B*s and *C*s. In what seems like an impossibility, I flunked art class in one of the eighth

---

2   National Aeronautics and Space Administration

grade terms due to my bad attitude. I wasn't belligerent or disrespectful to the teacher, and I was artistically capable to a reasonable degree, but was just resistant to what I considered frivolity at the time, and to the whole notion of being graded on something highly subjective.

Math teachers in the seventh grade made decisions at the end of the school year about which students should progress to *Algebra I* in the eighth grade, and which should continue with general math. I was one of the students not selected for *Algebra I*. This was an emotional setback. I was a bit angry knowing that I was capable, but I also had my confidence shaken at the same time. In hindsight, my teacher made the right decision since I didn't study very hard, was inconsistent in doing my homework, and my grades reflected this accordingly. But I was still a curious kid and loved school and learning, including math.

Extracurricular-wise, I learned how to play the trumpet, was in the junior high school band for two years and was recruited to play in the high school marching band—while still in junior high school—since the high school band was in desperate need for players. I gave up trumpet with some regrets to focus more on sports and part-time jobs. For sports, I tried out for the junior high school basketball team but didn't make it so I joined my church league for a season, whose standards were a little more forgiving, but quit after a month or so since the coach didn't play me much. (I wasn't that good, but I liked to play; I was *OK*. The coach played his stars.) In the spring of eighth grade, I joined the track team where, again, I did not excel at any event, but started a running routine that I've kept up with to this day (47 years later!). Outside of school and organized athletics, my friends and I engaged in seemingly every conceivable sport out there including basketball, football, ice and street hockey, tennis, swimming, to name a few.

Besides school and sports, I worked my first real job in the eighth grade as a paperboy, where I discovered that I liked work and business, since in those days paperboys essentially ran their own little businesses in serving customers, bookkeeping, and answering to the main office— all independently. In addition, I was always attracted to the military, and it was in the eighth grade at age 13 when I decided that, of the four branches of the military, I wanted to join the United States (US) Marine Corps once I completed high school.

High School, 1976-1979. High school covered tenth, eleventh, and twelfth grades in my town at the time. I was a normal teenager, relatively, in that I was overconfident and knew everything, experienced emotional highs and lows, made mistakes aplenty, and craved independence. I was curious and enthusiastic inside and outside of school. I couldn't wait to graduate, join the Marines, see the world, and experience real, nonfrivolous, purposeful work and adventure. I had many friends, where some were academically-oriented and others not. Most were good influences, but not all. With sports, I was my normal, mediocre self. I would have loved to have been on the varsity basketball team, but I didn't try out this time knowing the futility. Nonetheless I did sign up again with my church league's basketball team with the same outcome as before. (I quit after a couple of months given the same coach's tendency to play his favorites. The team overall was not that good.) Hence, I joined the track team, cross country team, and gymnastics team, where I was never the best but held my own. I became a *tri-captain* of the gymnastics team in my senior year (voting for co-captains by the team ended up in a tie, so the compromise was tri-captains), and my coaches presented me with a first-time, nontraditional trophy that I am very proud of today: "Best Attitude". Like a bonehead, I threw it away when I was in the Marines thinking at the time that it had no real meaning. I regret this tremendously.

Employment in my high school years consisted of pumping gas and performing minor car repairs (for example, changing tires) at my friend's father's gas station for a year, and later working as a cashier at a convenience store. As before, I enjoyed working, earning my own money, learning about the work world, and gaining valuable people skills by dealing with the public.

I was a poor student in high school—*kind of.* I liked my courses and teachers, and always enjoyed history. My two favorite courses, taken in my senior year, were physics and woodworking. I took a number of shop courses and loved them all. I enjoyed working with my hands, and the skills I learned in these shop courses continue to serve me to this day. Admittedly, I struggled in appreciating poetry required for English class, and rebelled somewhat like I did during junior high school art class.

I paid attention in my classes and wanted to do well. However, I performed unsatisfactorily, not because I couldn't do the work, but because I *didn't* do the work, even though deep down I wanted to. I actually enjoyed doing my homework when I did it. I just didn't do it a good portion of the time and my grades reflected this. I didn't know how to study properly, and didn't know how the learning process worked until it was too late. When I did pass in my homework on time, I felt a true sense of satisfaction and resolved to keep doing it, but my resolution faltered after I arrived home, went to work, attended sports practice, turned on the TV, or went out with my friends. Unfortunately, when I did do my homework, I did not put in my hardest effort most of the time.

As with many things in life, there's more than one reason to do or not do something. One reason for me not doing my homework regularly was FOMO[3], where the thought was that if I did my homework as I should have, I would have less time for friends and socializing. Another legitimate reason was that I was not guided academically at home. I had a good home life, but my parents never checked to see if I was doing my homework as I should have, and probably wouldn't have helped much anyway due to their limited education. I did not know the mistakes I was making at the time, not just with math but with other subjects and my attitude and approach towards them. I had trouble distinguishing between hard physical labor—which I demonstrated—and mental work, and took too much for granted. I took some

---

3   Fear of Missing Out

advanced placement classes (though they didn't call them that at the time), but fared poorly, not because of disinterest, but because of a lack of the hard mental labor required. I watched too much TV (a hypnotic and insidious device), procrastinated, crammed for exams, and irresponsibly consulted my academically astute friends the night before.

To complicate things, I didn't see the relevance of subjects such as advanced math, especially when I was overly focused, overconfident, and idealistic on joining the Marines and thinking that these subjects were not important in the military. I wanted to serve my country, experience some travel and adventure, and maybe even be part of history—consequences be damned. I envisioned a 30-year career with the Marines at the time but was also realistic that this may not work out after a first four-year enlistment. Had a teacher or tutor tried to intervene at the time regarding my academics and misaligned priorities—and some tried directly or subtly—his or her attempt would not have done any good; I wouldn't have listened unless he or she was an ex-Marine or ex-military.

Put all this together and the result was an unbalanced focus, lack of academic confidence, and self-fulfilling prophecy reflected with grades of Cs, Ds, and a few Fs as high school progressed into my senior year. I did apply to the US Naval Academy at Annapolis, Maryland (sometimes referred to simply as "Annapolis") with an intention of choosing the Marine Corps option, but I was not accepted. Had I been an admissions official at Annapolis, I wouldn't have accepted me at that time either due to my grades.

Though not the only reason, the beginning of my academic decline in my adolescent years seems to have started during that aforementioned seventh-to-eighth grade algebra-transition issue, where I self-decreed that I was not smart enough to become an engineer, scientist, or astronaut, and concluded that academics were irrelevant anyway since I was going to join the Marines (Priority 1!). How wrong I was.

US Marine Corps, Active Duty, Enlistment #1, August 1979 – August 1983. I graduated from high school in June 1979 at age 17. I worked my cashier job and had a lot of fun that summer, before arriving at Parris Island, South Carolina at the end of August for Marine Corps boot camp. Upon boot camp graduation that November, I returned home for ten days of leave before reporting to my first duty station at Naval Air Station (NAS) Memphis (called "NAS Memphis" even though it was located in nearby Millington, Tennessee). I turned 18 roughly two weeks after my arrival there. NAS Memphis was where the Navy and Marine Corps sent all their aviation mechanics and technicians for schooling to learn how to fix and maintain their aircraft. Upon graduation, the newly minted aviation mechanics and technicians were sent out to the "fleet" (formally called the "Fleet Marine Force" (FMF) for the Marines. NAS Memphis's mission has since been relocated to NAS Pensacola, Florida.)

My stint at NAS Memphis lasted approximately seven months where I attended three schools (~courses)—*Aviation Fundamentals, Basic Electricity & Electronics,* and *Avionics Technician Course*—to become an avionics technician, my primary Military Occupational Specialty (MOS). (In case you didn't know, *avionics* is simply the merger of *aviation* and *electronics*. An avionics technician troubleshoots, repairs, and maintains electrical and electronic systems on aircraft.) "Barracks Support", which essentially consisted of all-day cleaning and maintenance work details, filled in for those time periods while waiting for classes in the respective schools to begin. My time as a student at NAS Memphis was an eye opener on the subject of mathematics. I started to respect higher level math and see its real-world applications, not just to electronics but everywhere. This is where math started to "click in".

After graduating from NAS Memphis, I received orders to my first aviation assignment to the Marine Corps Air Station El Toro (MCAS El Toro; now closed) outside Irvine, California, where two days later, I was sent to my first FMF squadron at Marine Corps Base Camp Pendleton, California. My job there was to troubleshoot, repair, and maintain electrical and electronic equipment on the OV-10 Bronco aircraft.

Additional revelations came to me in the FMF where I witnessed further applications of mathematics to science, technology, and the wonders of engineering. While working on and around aircraft I had to ask myself many times:

- *How did they manufacture that component?*
- *How did they figure that out?*
- *Who planned out this complicated, well-organized, efficient, effective airfield and air operations?*

- *How did they organize the engineering and manufacturing that was needed to complete this huge, highly sophisticated, elegant piece of machinery?*
- *Why was this aircraft feature designed and configured the way it was?*
- *Why did they choose this particular material for this part of the aircraft over others?*

As many in the military do, I became somewhat disillusioned with the Marine Corps as time went on. I was, and still am, an idealist of the what the Marine Corps should have been versus what it really was. A significant time passed before I came to find out why things were the way they were. Knowledge of history helped. During this time of disillusionment, joined with multiple periods of anger, frustration, long work hours, and interpersonal strife, there were significant moments available for thinking since there's a lot of *waiting* in the military. I must add that my time in the Marines was certainly not all negative. In fact, those years were mostly positive, and I consider them to be the best years of my life, though imperfectly so. I learned a lot from my superiors and fellow Marines—technically, culturally, racially, ethnically, good, bad, and otherwise. The Marines, and military in general, is comprised of people from all over the country and from every conceivable background. Many of my preconceived notions and stereotypes about people and places were dispelled. I traveled a lot domestically and internationally, and experienced the adventure I signed up for. I made plenty of mistakes but gained valuable insight from them (most of the time). I was exposed to bad influences, rough crowds (a few probable future convicts), troublesome situations, and yet was able to keep my nose relatively clean. Many good officers and senior noncommissioned officers served as mentors and cut me a lot of slack when I screwed up. Due to seemingly constant work demands, irregular hours, and travel obligations on this first four-year enlistment, there was little time for night school or any kind of extracurricular activities that required a regular commitment.

Upon discharge as a sergeant at age 21 and some reflection, I felt like I understood the larger world, had a sense of purpose and direction, and realized that I had built up a lot of *steam* over the past four years. I was proud about what I had accomplished and that I kept my cool, since I could look upon my first enlistment in the Marine Corps as a four-year anger management course. I gained a great deal of technical experience, knowledge, and people skills that I never would have acquired otherwise. At the same time that I had this relative wisdom, confidence, and maturity about the world, I still had some unchecked overconfidence and judgmental issues. Neurons in my brain were still connecting. Deep down, with all the Marine Corps' flaws, I was glad I joined and served as an avionics technician, learned (mostly) from my mistakes, enjoyed the camaraderie, and made some lifelong friends. Now it was time to leave active duty, get some formal schooling, and learn what civilian life had to offer.

Gap Year, September 1983 – May 1984. Upon discharge from active duty, I immediately returned to my home town, moved back in with my parents due to limited funds, painted my parents' house, paid them $300 in monthly rent, landed a part-time job, became a full-time student at Bunker Hill Community College (BHCC) in Boston, Massachusetts, and joined the Marine Corps Reserve as an avionics technician again, this time working on UH-1N Huey helicopters at NAS South Weymouth (another base that has since closed), situated in South Weymouth, Massachusetts. Although not large, the extra income from the one-weekend-per month obligation with the Marine Corps Reserve was necessary for this period of part-time employment and full-time student expenses.

As a third source of income, I started to draw on my veteran's educational benefits from the Veterans Education Assistance Program (VEAP), which, at the time, was significantly constrained compared to what veterans receive today. VEAP came in between the Vietnam Era GI Bill and the more recent GI Bill versions. VEAP was inferior to all of them, where with VEAP one could obtain a maximum of $8,100. One third of the money, $2,700, was contributed by me, and the US Government kicked in the $5,400 difference. I wasn't complaining, but even in the 1980s, this didn't amount to much considering tuition, books, fees, food, shelter, clothing, and transportation. Fortunately, I wasn't paying a college directly for room and board.

My educational goal at the time was to earn an associate's degree in electronics technology in two years (the electronics training I received in the Marines was good but incomplete), then obtain what is called an Airframe and Powerplant (A&P) license from the Federal Aviation Administration by attending an A&P-accredited school for another year and a half, and subsequently taking and passing an exam, so that I could work on passenger aircraft at an airline as a career. I learned from Marine Corps active duty that documentation matters such that I knew enough that relying solely on my military experience working on aircraft was insufficient to work at the airlines. I needed substantiation in the form of a formal education backed up by a corresponding degree and license.

Before starting my curriculum at BHCC, I had to take a math proficiency exam. I was relatively confident going in even though I hadn't used any algebra since my time at NAS Memphis, but that confidence took a blow when I received the results. I didn't have enough competence in algebra to begin the electronics courses and had to take a course called *Foundations of Algebra*—the equivalent of ninth grade *Algebra*. Remedial math! I had to essentially start all over again!! Since I couldn't start my electronics courses right away, I took *Foundations of Algebra* in my first semester, and *College Algebra* in my second, along with my general education courses in parallel (such as college English, history, foreign language). I studied with direction and purpose this time around, worked hard at balancing work and school, and earned all *A*s at BHCC except for a *B* in *Intermediate French*. My study habits were imperfect (I was still learning how to study both efficiently and effectively) but they did improve over this introductory year of college.

During this year at BHCC with my improved study habits and academic work ethic, I had a revelation in that I was smart enough to become an engineer. But at the same time, I was itchy for more travel and adventure, missed the active duty Marine Corps in many ways, and realized I was still relatively young at age 22. I also knew that I needed more math—and money—to begin a formal engineering curriculum. So, I came up with a new plan: Enlist in the Marine Corps for another four years, volunteer for embassy duty to satisfy my adventurous appetite, and continue to go to school part time to bring my math skills up to a freshman-engineering curriculum level, all the while saving money for more schooling.

US Marine Corps, Active Duty, Enlistment #2, June 1984 – August 1988. That one year of education at BHCC and in the Marine Corps Reserve made a big difference in my way of thinking, in general, and in my approach to the Marine Corps. The importance and relevance of a formal education became even more apparent. I became much more discretionary in selecting friends and avoiding bad influences.

The Marine Corps assigned me to be an avionics technician again, but on a different aircraft, namely, the A-6E Intruder aircraft at MCAS El Toro. I liked my job very much for the most part, working it for one and a half years, which luckily consisted of relatively regular work hours. This allowed me to execute my new plan and become a part-time student at Saddleback College (a local community college) at its campuses in Mission Viejo and Irvine, California. I jumped at this opportunity and took courses in *Trigonometry*, *Precalculus*, *Mechanical Drafting*, a very valuable one-credit course called *Introduction to Engineering*, among some other electives. After almost one year at MCAS El Toro, a team of Marines from the *Marine Security Guard (MSG) Battalion* based at the Marine Corps Development and Education Command (MCDEC) in Quantico, Virginia came to our base recruiting for embassy Marines. I attended their presentation, later applied to become an embassy Marine, was accepted to the program, took on a secondary MOS as an embassy guard, and was given orders to report to the MSG Battalion at Quantico in December 1985 for embassy guard training.

MSG School, as it was called, was an intensive six-week curriculum comprised of everything from side-handle baton techniques to State Department organization, repeated inspections, table manners, espionage practices (to avoid being exploited by the bad guys), food sanitation, terrorist methods, weapons training, cultural norms and sensitivity, and many security-related duties. I progressed through *MSG School* satisfactorily, and was assigned to my first post at the US Embassy in Bridgetown, Barbados, which lasted approximately thirteen months. A twelve-month tour at the US Embassy in Warsaw, Poland followed. (Poland was a communist country at the time.) Two months at the US Embassy in Paris, France trailed Warsaw.

Embassy Duty was an exposure to: the US State Department organization and operations; foreign cultures; the immigration process; consulates; foreign policy in action; diplomats; US Peace Corps; US Agency for International Development; other US Government and international agencies; host countries' governments and their functioning, as well as those of other countries and their embassies. Perhaps the most valuable part of embassy duty was getting to know the locals and seeing the world from their perspective—culturally, racially, politically, socially, economically, and so on. Of my three embassy assignments, I enjoyed Barbados the most, not because of the tropical weather and palm trees, but due to the warm, kind, and friendly people and friends that I came to know there.

Once my time in Paris ended, I was sent back to the Marine base at Quantico where I was honorably discharged from my second four years of active duty at age 26. I had a relatively sound financial plan in place for school with roughly $25,000 in assets and what remained from my VEAP account in educational benefits.

While serving in Barbados, I took a correspondence course managed by snail mail letters (email for most people was a notional concept at the time) in first-semester *Calculus* from the University of California at Berkeley's Extension School. While in Warsaw, I did the same but for *Introductory Chemistry* (non-laboratory). I also applied, and was accepted, for the Fall 1988 semester at the University of Massachusetts at Boston (UMass Boston). Upon discharge from Quantico, I moved back to the Boston area, moved back in with my parents for two months until I could find an apartment, joined the Marine Corps Reserve again at NAS South Weymouth for a two-year enlistment as an avionics technician working on A-4M Skyhawk aircraft, found a part-time job, and started my formal, full-time engineering curriculum at UMass Boston. I was laser-focused and academically ready for this field of study now that my math and chemistry skills were where they should be for a freshman undergraduate.

Undergraduate, UMass Boston and UMass Amherst, September 1988 – May 1992. By this time, I had a good career path to become an engineer, and definitively decided on majoring in mechanical engineering over electrical engineering by the end of my freshman year. (The two curricula were essentially the same in the freshman year.) I transferred to UMass Amherst for my junior and senior years since UMass Boston did not have a formal four-year mechanical engineering curriculum at the time (just the first two years). I did not reenlist in the Marine Corps Reserve soon after starting classes at UMass Amherst due to academic commitments, a much longer commute to NAS South Weymouth, and discontentedness with this second reserve unit; it was the right decision. I lived frugally and drove a very unglamorous 1985 Chevrolet Chevette that I paid $1.00 for to my father, who paid $1.00 to his company for it before my anticipated military discharge. (The vehicle was the company's delivery car that the company was getting rid of. This $1.00 car lasted me for five years.)

I slowly went broke during my time at UMass Boston and UMass Amherst and progressed into debt. My $25,000 in savings and VEAP account were exhausted, so I took out three US Government loans, and also borrowed money from my mother and one of my sisters. If all I had to worry about were tuition and fees, I would have been covered, but I was also paying for food, rent, clothing, car maintenance, insurance, fuel, and other expenses as an undergraduate.

I was not a student genius, but was considered a hard, focused worker. I graduated from UMass Amherst at age 30 with a Bachelor of Science degree in mechanical engineering in May 1992, owing approximately $25,000 from my student loans (yes, a coincidental reverse in financial status) and personal loans to my mother and sister, which was significant in 1992. Thankfully, I had a mechanical engineering position lined up at an aerospace and defense firm upon graduation.

Since Undergraduate, June 1992 – Present. I was able to pay off my debts in approximately two and a half years. My lifestyle was very frugal. I lived like a pauper (relatively) with a cheap car, cheap studio apartment, and austere overall lifestyle until my debt was paid off. I was then able to start regularly saving and investing a reasonable amount of money.

The investment in education and math paid off. I now have 30 years of engineering experience working at three different aerospace and defense companies, and my financial position has noticeably increased. Academically, I took advantage of my employers' educational benefits and earned, via night school, a master's degree in mechanical engineering from Rensselaer Polytechnic Institute, and a master's degree in business administration (MBA) from the University of Delaware. From there, I acquired *Professional Engineer* licenses in Delaware and Massachusetts. Of the math courses required for my bachelor's and master's degrees, I tried to learn the subject matter thoroughly, apply the math to my work (directly or indirectly), and fully understand their phenomenal implications and usefulness.

I still continue with informal tutoring of middle and high school students where math tends to be the main focus of the students' homework troubles. These students keep me on my toes mathematically and otherwise. In my mind at age 61, my journey is nowhere near the end. I have many retirement plans, where *retirement* does not mean to stop working.

> *The secret to success is constancy to purpose.*
> - Benjamin Disraeli, British prime minister, statesman, and novelist, 1804-1881

Conclusion

Now that I've identified you and established myself with some mathematical "street cred" to some extent, let's conclude this introduction and move on to the book's core.

This book will hopefully help you learn from my limited wisdom, not just for your improved grades but for direction, purpose, resolve, thorough understanding, and right attitude for the future pursuit of mathematics—in school and out. The book is meant to be preachy, but not in a way so as to be abrasive, alienating, or annoying. I hope it will keep your interest and will be, if nothing else, somewhat entertaining.

2

What is Math?

Mathematics is the tool specially suited for dealing with abstract concepts of any kind, and there is no limit to its power in this field.[1]

- Paul Dirac, British theoretical physicist, 1902-1984

Ask one hundred people what math is, including math teachers, and you're likely to get one hundred different, incomplete answers. Math is a branch of knowledge that you need to understand and fully function in our very interdependent, interconnected, and economic society. Some people are better at using it than others. Some need more in-depth, advanced mathematical knowledge in their careers. Nevertheless, we all need it to navigate and to understand the world we live in. Math is about much more than balancing your checkbook, getting the correct change at the store, tracking baseball averages, manipulating numbers in a spreadsheet, showing up for work on time, or making you a "more logical" and "better" person. So, what is it, and who cares?

Let's first look at some authoritative definitions and excerpts (there are many others), and then I'll complement

---

1   Courtesy of the Florida State University Libraries, Special Collections and Archives

them in a way in which I wish someone had told me when I was your age.

Definitions and Excerpts

The Random House College Dictionary, Revised Edition, © 1982[2]

- The systematic treatment of magnitudes, relationships between figures and forms, and relations between quantities expressed symbolically.

McGraw-Hill Encyclopedia of Science and Technology, 11th Edition, © 2012[3]

- Mathematics is not subordinate to natural science by being a handmaiden of it, and one can practice competently meaningful mathematics without being concerned with science at all.
- Mathematics is an indispensable medium by which and within which science expresses, formulates, continues, and communicates itself.
- [Its purpose is to] not only specify, clarify, and make rigorously workable concepts and laws of science, but also at crucial instances becomes an indispensable constituent of their creation and emergence.

---

2   The Random House College Dictionary, Revised Edition, © 1982. Used by permission of Penguin Random House LLC. All rights reserved.

3   Used with permission of McGraw-Hill LLC, from McGraw-Hill Encyclopedia of Science and Technology, Mathematics, Salomon Bochner, Volume 10, 11th Edition, © 2012; permission conveyed through Copyright Clearance Center, Inc.

Mathematics Dictionary, 5th Edition, James and James, Chapman & Hall, © 1992[4]

- The logical study of shape, arrangement, quantity, and many related concepts. Mathematics is often divided into three fields: algebra, analysis, and geometry. However, no clear divisions can be made, since these three branches have become thoroughly intermingled.
- Applied Mathematics – The [mathematical] study of physical, biological, and sociological worlds.
- Mathematics of Finance – The study of the mathematical practice in brokerage, banking, and insurance.
- Pure Mathematics – The study and development of the principles of mathematics for their own sake and possible future usefulness, rather than for immediate usefulness in other fields of science or knowledge. The study of mathematics independently of experience in other scholarly disciplines. Often the study of problems in applied mathematics leads to new developments in pure mathematics, and theories developed as pure mathematics often find applications later. Thus no sharp line can be drawn between applied and pure mathematics.

The New Encyclopedia Britannica, Volume 7, © 2005

- The science of structure, order, and relation that has evolved from elemental practices of counting,

---

4   Cengage Learning, Inc. Reproduced by permission. www.cengage.com/permissions

measuring, and describing the shapes of objects. It deals with logical reasoning and quantitative calculation, and its development has involved an increasing degree of idealization and abstraction of its subject matter.

Mathematics: An Illustrated History of Numbers, Edited by Tom Jackson, Shelter Harbor Press, © 2012, Page 8

- Math can be applied to anything—with varying degrees of success—and so defining it in terms of its applications results in a lot of confusion. It is like explaining the telephone by reading out of the phone book. Even a theoretical basis for dividing up math requires a lot of compromise due to the huge potential for crossover between subjects. In the simplest terms math tackles quantities, essentially different ways of counting; number structures, the patterns and linkages within; space, the characteristics of shapes and surfaces; and finally understanding change by tracing dynamic systems from instant to instant.

Supplemental Definition

Note that the definitions above are different from each other, and yet, not surprisingly, they somewhat converge, almost saying the same thing. Books, videos, teachers, professors, and so on will dance around the same theme. If you research definitions on psychology or baseball, you'll get different-but-similar results. The definitions above are helpful, accurate, and necessary, but for someone like me they are insufficient and incomplete. I want to see the purpose and application of something before spending countless hours sitting in class, doing homework, and studying for exams. If I've described you adequately in the *Introduction*, you feel the same way. For other subjects (for the most part), it doesn't matter: We get it! But math is where the capable but incredulous get stumped, which is what this book sets out to cure.

I won't be so arrogant, overconfident, and irresponsible to substitute for the experts, but I will offer my own supplemental definition:

- Mathematics is a language and set of tools that is necessary—in many cases, nonobvious ways— to understand, explain, and function in science, engineering, government, construction, trades, business, agriculture, economics, finance, insurance, and innumerable other places, by allowing us or helping us solve real-world problems and make things work, and to think and communicate critically, analytically, comparatively, qualitatively, and quantifiably to make sound, balanced, knowledge-based decisions and judgments.

Language

My *Precalculus* professor at Saddleback College (a part-time math professor and full-time physicist) was the first person who revealed the notion of mathematics being a language. This was another "click-in" moment for me. He's not the only one to have come up with this language concept. (Note Galileo's quote later in this chapter.) Many others have done the same, including the comparison to math being a set of tools. The problem for most middle and high school students is that these definitions and complements thereof, are not always communicated, or done so clearly and emphatically as they should be.

To help explain the language part, let's use an analogy via text. Letters, characters, symbols, phonics, words, punctuation, grammar, and structure all become sentences, paragraphs, and the like. They, in turn, combine to form:

- Instructions
- Correspondences
- News
- Advertisements
- Plays

- Books
- Statutes
- Contracts
- Constitutions
- Warnings

An endless list.

For math, the situation is similar but for numbers, graphs, shapes, formulas, symbols, patterns, and so on, which then help construct:

- Products
- Services
- Processes
- Systems
- Hypotheses

- Tools
- Foods
- Software
- Medicines
- Theories

- Engineering and
  scientific principles

- Infrastructure
  (bridges, roads, water
  supply)

The list goes on unendingly. So, like text, which is also an abstract concept, the language of mathematics can be used by we humans to profound effect. As for the set-of-tools concept, see the chapter *What Do You Do with All This Stuff?*

History Overview

Being who you are, you want to know more; you don't just accept things without some thought. I'm sure you've asked yourself questions during math class or in doing your homework such as:

- *Where does this come from?*
- *Who thought of this?*
- *Why does it exist?*

I make no pretense posing as a historian, but I have done some research into the history of mathematics, which I have found to be both humbling and fascinating. As with any research, I've learned that I only scratched the surface. There's an enormous number of artifacts and information on math history.

So, where does math come from and why does it exist? If for no other reason, it was, and still is, needed for everyday life and civil administration. Excluding the curiosity and entertainment factors for now, and focusing on the practical applications in ancient times, math was needed for counting, surveying, fractions, making measurements of all kinds, defining and organizing dates and times, navigation, commerce, logic, reasoning, money, and abstractions that may or may not have had immediate applications but did later. Early developers had to apply this language and these tools in these fields to help explain, quantify, qualify, describe, estimate, construct, and so forth, to function for such instances as:

- *I want that pig, but I don't want that blanket in return.*
- *What parcel of land is yours, and which one is mine?*
- *I don't want a whole hekat[5], just half of one.*
- *Make it 10 cubits[6] high.*
- *We want to sail from Alexandria to Athens.*
- *I estimate that 1,000 soldiers will be needed to defend our city from the enemy.*

Development. Math developed gradually. Its evolution was anything but straightforward and was rife with errors (not to be confused with mistakes), but valuable, nonetheless. One step or milestone was built on another via a combination of creativity, intelligence, and *borrowed brilliance* throughout history, and still is. As Isaac Newton[7] famously said, "*If I have seen further, it is by standing upon the shoulders of giants.*"

Math is a composition of contributions from many different cultures, civilizations, and individuals resulting from such things as trade, shifting borders, military conquests,

---

5   Ancient Egyptian unit of volume
6   Ancient unit of length
7   British physicist and mathematician, 1643-1727

empire fluctuations, economic competition, internal and external collaboration within or between nations, and general necessity. For instance, as the fields of astronomy and physics matured, they increased the need to understand and express the infinitesimally small and infinitely large, and to explain and comprehend the physical world such as time, mechanics, optics, heat, and electricity. The curiosity and entertainment factors were germane. On top of practical applications, mathematics may or may not have had an impact on a particular problem that needed to be solved.

> *I must study Politicks [sic] and War that my sons may have liberty to study Mathematicks [sic] and Philosophy.*
> - John Adams, 2nd president of the US, 1735-1826

Humbling and Fascinating. Whether learning about early geometry and the Pythagorean theorem from other non-Greek ancient cultures (yes, it predated Pythagoras[8]) and its further advances by the ancient Greeks, the founding of algebra from Islamist mathematicians, of logarithms developed by John Napier[9], or Carl Friedrich Gauss's[10] mathematical abilities, I became acutely aware of my own limitations and incomplete math knowledge. I go from a feeling of cleverness to one of humility quickly. At the same time, I can't help but become fascinated. The fascination

---

8   Greek philosopher and mathematician, circa 570-490 Before the Common Era (BCE)

9   Scottish mathematician and theological writer, 1550-1617

10   German mathematician, 1777-1855

relates not only to the technical awe, but also the human trials, tribulations, intrigue, and rivalries involved. The rivalry between Isaac Newton and Gottfried Leibniz[11] over the development of *Calculus* is a classic (they developed it independently), but not the only one, never mind the political and social contexts in the respective periods. My own love of history keeps my attention.

Invented or Discovered? They say that necessity is the mother of invention, but, like the sciences: Was math invented or discovered? This is an age-old question. If the number π is the ratio of a circle's circumference to its diameter, and somebodies figured this out (the search for π started with the ancient Babylonians and Egyptians; Archimedes[12] and many others refined its accuracy over many years), did they invent this or didn't it always exist? When mathematicians do their thing, are they creating something, or uncovering these tools and language artifacts that existed before the world and universe began? I recall one tutoring session where my student and I were discussing the relevance and applicability of π, where she had asked, "*Is this still true?*" My response was, "*It's been true for over 2,000 years.*" What my answer should have been was, "*It has always been true, and always will be.*" Give this some thought.

---

11   German philosopher, mathematician, and political adviser, 1646-1716
12   Greek mathematician, circa 287-211 BCE

Conclusion

In this chapter, I've supplemented formal math definitions with the proposition that math is a language and set of tools that is necessary for all of us in daily survival, even though we take it for granted.

> *Philosophy [nature] is written in that great book which ever is before our eyes—I mean the universe—but we cannot understand it if we do not first learn the language and grasp the symbols in which it is written. The book is written in mathematical language, and the symbols are triangles, circles, and other geometrical features, without whose help it is impossible to comprehend a single word of it; without which one wanders in vain through a dark labyrinth.*
>
> - Galileo Galilei, Italian scientist, astronomer, and mathematician, 1564-1642

Math didn't come by accident. It has been developed over the centuries by numerous contributors for a variety of reasons in indirect, iterative, and nonlinear ways. Dig deeper into the history of mathematics, and you will start to appreciate, and—perhaps like me—find out that we're not as smart as we think we are. Expect to be captivated, nonetheless. I've also been astounded, humbled, and outdone many times by the mathematical abilities of middle and high school students that I've worked with.

Math is a critical subject. It doesn't end. We will always have a need for it. Academic pursuit and research into mathematics will continue. You will never see the full applications of math, even if you pursue a career in engineering, theoretical physics, and related fields. None of us will. (We will never see the full application of any field of study when you get right down to it.) But know that this language is always there facilitating your life, work, and world. We can't function without it. And, like a handyman, you will always have a need to reach into your toolbox for a particular tool to solve a problem at hand.

Now let's move on the next chapter to see what is done with it.

3

What Do You Do with All This Stuff?

If in other sciences we should arrive at certainty without doubt and truth without error, it behooves us to place the foundations of knowledge in mathematics.
- Sir Francis Bacon, British author, lawyer, statesman, and philosopher, 1561-1626

At the end of one of my *Ordinary Differential Equations* classes at UMass Boston where students from my math class were leaving, and students from the incoming replacement class were arriving (it may have been a social science class), one of the incoming mystified students asked me, after looking at the chalkboard where my math professor had left his writings, *"What do you do with all this stuff?"* This was a perfectly legitimate question, considering all the abstract scribbles. Unfortunately, I couldn't definitively provide an answer since I was still learning the applications. Yet I knew the right answer was powerful and profound. I've heard this question every so often during tutoring sessions, and as I've already related, I asked myself the same question when I was in high school.

Knowing what I know now, the quick answer to the incoming student's question is that the math is (yes, including *Ordinary Differential Equations*) used by regular people every day to help solve real-world problems and make things work—directly or indirectly. But as we learned from the definitions, there's a direct and indirect correlation to science, engineering, and other fields, and sometimes not at all. Recall that math can be performed just for its own sake, as in pure mathematics, with no immediate applications.

Many middle and high school math teachers as well as other professionals don't know themselves what to do with much of it, especially "advanced math". Similarly, they may not know why certain aspects of math are being taught. An example of this is *imaginary numbers*. I had a high school math teacher once ask me—when I was a working as an engineer—what they're used for (even though she taught math). Physicists may not know how to apply a particular mathematical technique or principle until the need is realized. We're all limited to our degree of education, knowledge, capabilities, and experience with mathematical applications.

Solving Real-world Problems and Making Things Work

For those people who are solving real-world problems and making things work with math directly, what math do they use? I'll give you the classic, frustrating answer: It depends. They use whatever is appropriate. They go into their math toolbox and decide based on their knowledge and experience with math.

An Analogy. Let me answer the question in a different way using an analogy. A professional car mechanic has an immense set of tools at his disposal to repair and maintain cars. Which tool does he use to conduct his work? All of them? Some of them? Only one of them? It depends. He uses whatever is pertinent for a specific application, and his decision is based on his knowledge and experience. He certainly doesn't use all of them for every problem. In fact, he may not know what to do with some of the tools that he has in his possession. If he has never repaired an automatic transmission before, he may not know the purpose of a universal spring compressor and how to use it properly, even though he knows it's important and may need it in the future. You can then expand on this theme and apply the analogy to special tools used in a hospital emergency room, kitchen, computer, wood shop, and artist's loft. Math is needed to communicate the scientific, engineering, finance, economic, and marketing aspects of our world's problems. Without this language and set of tools, we're dead in the water. But let's be clear: just like our mechanic with his set of tools, math in and of itself is not a cure-all. Let's broaden our car mechanic analogy further.

Before proceeding on a particular job, the mechanic evaluates the problem at hand cautiously, uses basic car knowledge, physics, chemistry, and experience, and takes into consideration all other relevant factors. That is, there are limits to everything and the mechanic still must "think", and not rely solely on his tools. Here's an example. A car owner keeps coming into his shop with the same problem: extensive, premature front brake wear. Yes, the mechanic can replace the brake pads and rotors, or turn the rotors, if necessary. But upon further questioning of the owner, the mechanic realizes that the owner drives with two feet—one foot on the gas pedal and the other on the brake. In other words, the problem is not the car or brake system, but the owner's faulty, self-defeating driving technique. The car owner is the problem. The mechanic advises the car owner not to drive this way anymore since it leads to premature—and expensive—brake wear. Common sense comes into play as well.

It's the same with math. A car, as an example, will not be designed solely with math. Nor will the design be strictly dependent on engineering (which is essentially valueless without math). Many other factors go into designing a car, and hundreds, if not thousands of people are involved. The other factors considered include: market demand; consumer preferences; costs for labor, manufacturing, materials, overhead; program management; manufacturing capacity; return on investment; supply chain; government regulations; and legal aspects; to only name a few. But the inherent complications associated with the other factors can't be remedied without communicating with the stakeholders in that necessary common language called mathematics.

What problems need to be solved? Well, where do we begin? Life is full of problems. We all have personal problems. Businesses have problems, as does any community of people. Local, state, national, and international problems abound. The military is rife with problems. The Earth we live on has problems aplenty (many that we humans created). Read on.

Indirect Use of Math. You may say, *"But I have plenty of problems and I make things work, and I do lots of things, and I* *don't use math with any of them."* Really? Are you sure? Let's think about that a little more deeply. You may—consciously or unconsciously   or may not use math directly, but you're certainly using it, benefiting from it, or using it *indirectly* by:

- Reading this book
- Getting dressed
- Making a phone call
- Driving a car
- Riding on a bus
- Playing video games
- Eating a hamburger
- Flushing a toilet
- Managing a bank account
- Walking on a sidewalk
- Getting a drink of water
- Stapling two pieces of paper together
- Looking at a clock
- Sailing a boat
- Receiving a paycheck
- Turning on a light
- Watching TV
- Baking a cake
- Standing on the floor
- Receiving a flu shot
- Participating in any sport
- Using email or social media
- Listening to a weather report
- Using a computer or calculator in any way

"Give me a break! How am I using math or benefiting from it indirectly by reading this book?" you might ask. Becoming blind to math, what it helps to produce, and taking its usefulness and applications for granted is easy. Here are some ways you are benefiting from math indirectly while sitting and reading this book. Consider:

- The chair you're sitting in, with its construction of wood, metal, or polymers (~plastics), and what went into its design, machinery that made it, extraction of raw materials from which it came, and businesses that made it available
- Your medical condition and the medical system that facilitated your health to allow you to see—never mind read—this book
- The agricultural and food distribution systems that are keeping you full and healthy enough to help you concentrate on reading and not on your next meal
- The chapters and pages that are numbered in a logical sequence
- The raw materials that made the paper and ink, and subsequent manufacturing processes that made the book
- The building or house that you may be sitting in along with its electrical, plumbing, heating, ventilation, and cooling systems designed to keep you comfortable
- Word processing software that was used to put thoughts and words "on paper"
- The monetary exchanges for book procurement made by you and other supply chain stakeholders in the background

You can broaden this approach to your student day to include the time to wake up and estimating the time to catch the bus or walk to school to get to your first class on schedule (where time is an abstract concept requiring math to measure and understand). Where does the list end, and not just with this book or with your day? It doesn't.

Using Advanced Math Directly. What about direct applications of some of that more advanced, "mysterious" math? For example, how are matrices applied to real life? In case you forgot or don't know what matrices are, they are those ordered, rectangular arrays of numbers, typically enclosed in parentheses or brackets, that are used as a shorthand method for organizing and managing only the essential data in a linear system of equations, which can then be added, subtracted, and have other operations performed with them[1]. Where they are used is not self-evident to the high school student, since they tend to be embedded in software, and are applied in such fields as:

- Fluid mechanics
- Electromagnetics
- Structures
- Astronomy
- Chemical processes

- Meteorology
- Computer science
- Heat transfer
- Biology
- Securities (stocks and bonds) analysis

Could matrices be used anywhere else? Of course, and they are used in many nonobvious ways. The same holds true in other facets of mathematics, including shapes and solids, many forms of coordinate systems, imaginary numbers, logarithms, and so on. Matrices and their associated operations are tools from our math toolbox that can be applied wherever

---

1   An example would certainly be illustrative here, but I want to keep my promise of not using any math.

and whenever appropriate to communicate in a language that couldn't be communicated in any other way to solve a particular problem. Compare this to the field of law where legal problems cannot be solved without communicating in a common language (for example, English) with agreed upon definitions.

Math on its own does not solve problems, but it is a critical tool in doing so. In addition to the science, engineering, and other math-intense fields, we also need to be practical, use common sense, and sound reasoning. When an analyst is performing a computer-based analysis in fluid mechanics, electromagnetics, chemical reactions, or finance, she is not simply pressing buttons and applying formulas to get a solid, definitive answer. (If that were the case, I and many other engineers would be much happier.) Integration of many technologies and other pieces of information—both quantitative and qualitative—are needed from other fields of knowledge, experience, and constraints to arrive at a final sophisticated solution. (And many times, that needed information is imperfect and/or incomplete.) Like our mechanic, we still have to *think*.

> *There is no branch of mathematics, however abstract, which may not be some day applied to phenomena of the real world.*
> - Nikolay Lobachevsky, Russian mathematician and a founder of non-Euclidean geometry, 1792-1856

Conclusion

Becoming blind to the applications and benefits of mathematics is natural. Keep in mind that when you're interfacing with a computer, cell phone, TV, or radio, these devices aren't working by magic. Clicking, swishing, scrolling, and manipulating software is relatively easy. What's going on in the background is quite sophisticated and math-dependent. Furthermore, using a calculator, which is far from a simple machine, is useless if you don't know what you're doing math-wise.

Without theory, we are lost. A common fallacy is that theory is useless; what matters is what happens in the "real world". This is bunk. The sounds that come out of a musical instrument will be nonsensical and sound horrible if the player has no musical theory. I recall taking sailing classes with many other would-be sailors. I noticed a remarkable difference between the students who paid attention in the classes and studied the sailing textbook, versus those that didn't and preferred to only learn on the water. The latter performed miserably (and unsafely) on the water since they had no theory or knowledge foundation on which to operate. As former President Eisenhower[2] once said, "*I tell this story to illustrate the truth of the statement I heard long ago in the Army: 'Plans are worthless, but planning is everything.'*" Replace *Plans* with *Theory* and you'll see my point.

---

2   Dwight D. Eisenhower, 34th president of the US, supreme commander of the Allied forces in western Europe during World War II, 1890-1969

Your life is just beginning, and your world will be rife with challenges that my generation, and those before mine, never had to face. These challenges will require math—and many other things—to solve and/or manage. As one of my physics professors once told me, "*You can never have too much math.*" This is just like our mechanic analogy, where a mechanic can never have too many automotive tools (figuratively) since he can encounter any number of unanticipated problems calling for both common and specific tools. And being more proficient in any language (in this case, math) will always be to your benefit. As with the perplexed student who asked me "*What do you do with all this stuff?*" when looking at the chalkboard filled with intricate symbols and equations, math can be complicated at times, and if it sometimes looks difficult to learn and manage, it can be. Conversely, many times math just looks ominous, when, in fact, it is quite simple or methodically feasible. With math, one thing builds upon on another. Getting an *A* in algebra, and then expecting to progress to the next math course without remembering anything in algebra is not possible; prerequisites matter.

Applying math to a real-world problems can be awkward at first. Just like applying drywall compound for wall repair, hitting a tennis ball with a tennis racket, blowing into a trumpet, turning a socket wrench, using a sewing machine, or painting with a paint brush can be awkward in first attempts. We gain confidence and judgment with repeated application, intent, conviction, direction, practice, experience, and perseverance, just like in sports, music, or any activity that requires skill and discipline. Self-talk that says, "*I can't do it"*, needs to change to *"I don't know how to do it, but I can learn."*

Math on its own will not solve anything. There is no substitute for human judgment. To solve problems, you still must employ your brain, think critically with a substantive perspective, and use common sense. In the software world, the term "GIGO" is used a lot, and applies to this discussion as well, where GIGO means *Garbage In, Garbage Out.* (Load garbage, as in bad or nonsensical data, into the computer and garbage comes out.) But with an adequate set of tools, which includes math, among many other subjects that you learn in school, at work, and in life, combined with judgment that comes with experience, so that you can recognize and filter out the garbage, there's virtually no limit to the problems that can be solved.

Now let's cover the *reasons* to study and learn math.

4

Why Study Math?

A word to the wise is enough.

- Benjamin Franklin, aka Poor Richard, American printer, publisher, author, statesman, inventor, scientist, etc., 1706-1790

Now that we've defined math and have a little more insight about what to do with it, the next logical questions to ask are:

- *Why should we study math, and continue to study it?*
- *Why is math so important?*

Let me first say that math is a critical field of study. It is part of a balanced education but is insufficient on its own to navigate through life. Of course, this doesn't answer the *Why* questions. There are three main, interrelated reasons to study math, which I'll cover in some detail. But towards the end of the chapter, I'll provide even more reasons and considerations.

Three Main Interrelated Reasons

1. The first reason for all of us to study and become proficient in mathematics is for national security (and I don't just mean this in the military/defense context). Our national security—never mind global

security—depends on its citizens to be math-competent. We need math-capable people to understand, create, and manage highly technical, complex things that are math-dependent (such as nuclear power plants and financial systems). We would have societal collapse otherwise. We can, therefore, look at studying and mastering foundational math as altruistic where *I'm doing my part*.

2. On a more practical level, you personally need to study math for individual survival. You will need to gain employment, generate a livable income, steer through life, and provide economic options.

3. Since we humans tend to *look out for Number 1*, here is the most selfish reason to study math and to get your attention: People will—intentionally or unintentionally—take advantage of your math deficiency, with you being the individual you know best, but you also as a:

- Business owner
- Medical patient
- Tenant
- Government official
- Tuition-paying student
- Soldier, Sailor, Airman, or Marine

- Employee
- Investor
- Manager
- News recipient
- Citizen, voter, and taxpayer
- Consumer, customer, or client

The list goes on and on. Society benefits from your selfish vigilance, and there's not necessarily a conflict with any of these three reasons. Now let's dive into a little more detail for each.

National Security

The US is said to be the freest and wealthiest nation on Earth as well as the most technically complex. To preserve our freedoms, maintain our quality of life, and sustain our economic competitiveness, we need our citizens to have at least a basic and balanced education, which includes a sufficient level of mathematical proficiency to deal with complexity. This can be achieved by providing the needed educational opportunities to our youth, exposing them to their occupational possibilities to allow them to pursue their goals and aspirations, and giving them economic incentives to do so—both positive (rewards) and negative (consequences). When we're all "on board" and working together, a synergistic effect results where the whole is greater than the sum of its parts.

An educated population can then perform constant, objective self-evaluations, make improvements, and solve the inevitable problems of society. A balanced education allows us to be self-reliant collectively and on an individual basis. A "balanced" education is up for interpretation and has been for a very long time. When I refer to it, I mean the one most Americans are familiar with given our current model of western education (history, English, science, civics, math) but also to include physical education, music, arts, and exposure to trades and/or shop classes where they are not limited solely to vocational schools. Our society does not just need doctors, nurses, scientists, engineers, and finance professionals. Having the supporting mechanics, technicians, and crafts people is critical for society to function properly. And for those who do not want to do that work as a full-time occupation, they benefit with mathematical knowledge to be somewhat self-reliant.

I fear that your generation and subsequent ones will inherit troubles that previous generations never conceived possible. A multitude of problems exist now, and more are coming. Before transitioning from the altruistic, national security reason to study math to more self-centered ones, I'll leave you with a few of Thomas Jefferson's[1] words:

- *"Above all things I hope the education of the common people will be attended to; convinced that on their good sense we may rely with the most security for the preservation of a due degree of liberty."*
- *"It is safer to have the whole people respectably enlightened than a few in a high state of science [knowledge] and the many in ignorance."*

Individual Survival

Part of what's needed for you to survive in our capitalistic-but-interdependent society includes that language and set of tools called mathematics. You will need an income via running your own business or traditional employment, and

---

1   Thomas Jefferson, 3rd president of the US, vice president, secretary of state, diplomatic minister, continental congressman, principal author of the Declaration of Independence, etc., 1743-1826

most businesses require employees who have an adequate level of mathematical skills. You will also need to manage your personal finances, and make sense of the news and world, in general. College is not necessary to survive, but a college degree can help you achieve your goals, depending on what your goals are. And a properly accredited college or university will require successful completion of a certain level of mathematics before it grants you a diploma.

With a Solid Foundation in Math. A solid foundation in math establishes the basis for many other important subjects such as science, engineering, agriculture, economics, finance, technology of all types, and the trades. Math on its own guarantees nothing, but it helps increase options for you—employment or otherwise—given the constraints of the local, national, and international economic landscapes, and your own personal limitations, talents, and goals. Math is a significant educational component and has been trivialized or ignored in the past by many as not being so important. Mathematical skills help and allow for: the understanding of science; abstract thinking and concepts; survival in emergency situations; logical reasoning in many ways; making measurements and estimations; formulating critical assessments (not to be confused with the easy, impulsive kind that is devoid of substance); to name a few.

Math skills are advantageous in many obvious and nonobvious ways to yourself, the business you might work in, the government body that you may work for and will depend on, or any other type of organization. You, in turn, will benefit with a reliable revenue stream whether you become an employee, business owner, contractor, or something else. If you benefit your organization, you will be able to survive, and hopefully thrive. Your math skills, among others, will translate into your employment marketability, improved decision making and judgment, and so on. Always keep in mind the phrase, *Knowledge is power*, not to mention independence and survival.

Without an Adequate Level of Math. Without an adequate level of math, your goals could be made more difficult. Life doesn't always go as planned. Living by your wits and subscribing to the ignorance-is-bliss mentality is no way to proceed, where ignorance (not to be misinterpreted as stupidity) can take many forms. Being confident and content are great things, until something goes awry, and you're left helpless. Catastrophes, storm damage effects, emergencies, dangerous situations, can often be avoided or minimized with the right knowledge either with math directly or that which math can facilitate. Preparation, constant vigilance, and adopting a goal of self-reliance will help you lessen your dependency on others. As Andy Grove, founder and former Chief Executive Officer and Chairman of Intel, famously said, "*Only the paranoid survive.*"[2]

To help get my point across, let me tell you what you don't want. You do not want to be out of work due to a lack of knowledge and skills, being helpless in emergencies, subjected to unsafe conditions, being financially destitute, unnecessarily awash in debt, or a victim of fraud. You don't want to be *caught*

---

2   Andrew S. Grove, *Only the Paranoid Survive: How to Exploit the Crisis Points that Challenge Every Company* (New York: Penguin Random House, 1999)

with your pants down and living paycheck-to-paycheck. You must anticipate the future with a reasonable and calculated probability of economic occurrence for the economy as a whole and for you personally. As you read this, the economy may be strong as reflected by a healthy job market with lots of opportunities. It won't last forever; economies tend to go through cycles. A time will come when there will be a lack of employment options, perhaps severely so. At the time of this writing, the unemployment rate is very low, but there's a wide gap between high- and low-paying jobs, where many high-paying jobs remain vacant since employers have a hard time finding employees with an adequate level of math and technical skills to fill them. Automation and other factors will continue to decrease employment opportunities in some areas and increase competition among peers, exacerbating the gap we're already experiencing between the so-called *haves* and the *have-nots*.

A college degree unto itself is no panacea for employment or self-sufficiency woes. Not long ago, the economy experienced high unemployment, yet many high-paying job openings remained empty simply due to a pool of unqualified candidates. There was a glut of college graduates who couldn't find a job that paid a sufficient amount of money to pay off student debt or to have the quality of life they had envisioned prior to college. The reasons vary from majoring in unmarketable fields or attending college for its own sake without passion or purpose. Many of the aforementioned high-paying job openings existed in the trades where no college degree was even required, but a significant amount of technical knowledge was.

What if you don't want to be an employee but rather an employer; that is, a senior manager within a company or a business owner? Your department, organization, or business may suffer unnecessarily due to your math deficiency. Industry is math-dependent, and not just for bookkeeping and accounting. Other math-dependent facets of business cover business development, marketing, engineering, research and development, operations, supply chain and logistics, human resources, to name a few. And you will need employees with a sufficient degree of mathematical and technical know-how to operate the business. During the hiring and vetting process of technical people, especially scientists and engineers, how will you know if they know what they are talking about, or if they are fooling you—and themselves in some cases? You will need enough mathematical knowledge to make the evaluations.

None of this may be obvious to you yet. Realizing the value of things—especially food, shelter, and clothing—is difficult until you've been out on your own with no financial backup and have to pay for them with money that you've earned and depend on. A math deficiency or ignorance, in general, makes steering around life's obstacles, of which there is an abundance, that much more difficult. I've experienced these obstacles and the related trouble more times than I'd like to mention, in instances where I should have known better, which leads us to the next section.

People Will Take Advantage of You

Like it or not, people can and will take advantage of you deliberately or inadvertently. This is a part of life. From a self-centered point of view, this is arguably the most important reason to study math. You have to *look out for Number 1*, not only to survive, but not to be exploited. By looking out for yourself (and perhaps your family later in life), society benefits indirectly. Society benefits even more—and you personally as well—when you help your neighbors, coworkers, citizens, and fellow students in their difficulties; in other words, when we all work together. As I mentioned, being taken advantage of can be inadvertent. Well, who would or could then, other than criminals and scammers? Here's a list of some of them who could potentially do so (and the list is far from exhaustive):

- Home contractors
- News pundits
- Salespeople
- Engineers and scientists
- Politicians and government officials

- Doctors, nurses, health care companies, and hospitals

- Family members
- Managers of utilities
- Teachers and professors
- Retailers and merchants
- Bankers, financial advisers, and financial institutions
- Employers, employees, coworkers, and colleagues

You can also fool yourself, and you can be guilty of deliberately or unintentionally taking advantage of these same people. (We're all hypocrites to some extent.) In what ways can you be taken advantage of? How about the following as starters?

- Personal finance and consumption of goods and services
- Public finance and government
- Health and well-being
- Business-to-business transactions

Math is essential in helping make sense of things to prevent being duped. But you may say, "*No one is going to take advantage of me! I'm too smart. I've been around and know how the world works!*" As I've said before, don't confuse stupidity with ignorance. You will always lack knowledge in some way and will forever be at the mercy of the various media sources like the news and advertisers, plus hypes, trends, fads, and TV shows. You will also be exposed to social media, as well as sloppy, incomplete, biased, and/or false news. A truly balanced education will allow you to distinguish between false news, unsubstantiated claims, conspiracy theories, slanted and subjective reporting, or sensationalized and emotionally based news, to get the real story. Your goal should be to have the ability to think for yourself and not be subjected to fallacious arguments made by "experts". (*They are considered experts, have made a name for themselves, and speak eloquently and persuasively. Therefore, they must know what they are talking about.* Maybe. Maybe not.) You do not want to be at a disadvantage when listening to these so-called experts, so you need to be able to recognize their ignorance (recall the unintentional part), or their deception, sophistry, conflicts of interest, or malfeasance in government, business, military, person-to-person situations, and so forth.

> *I for my part would prefer tongue-tied knowledge to ignorant loquacity.*
> - Cicero, Roman statesman, scholar, and writer, 106-43 BCE

Personal Finance and Consumption of Goods and Services. As a manager of your personal finances and a consumer, you can be taken advantage of in so many ways. You will pay for ignorance, and for not doing your homework, especially your math homework. I've been my own worst enemy on some occasions and can speak with painful experience here as well. This ignorance also covers money that could have been made on lost opportunities. Bear in mind the maxim that *you can't spend what you don't have.* Debt management—as opposed to wealth creation—is no way to live. Credit card companies' fine print and sneaky marketing methods can lead to your downfall if you fail to understand them fully. Loans and credit are things to avoid but are many times necessary evils. Banks are notorious in charging you "mystery" fees, which make up a significant source of their revenue. Not comprehending inflation and interest rates and their effect on you and the economy can leave you at a major disadvantage.

Many people have a dream of running their own business or owning a home. If you're one of them, you will need financing. When the lender cannot specifically explain where the numbers for a loan or mortgage come from and how they are calculated, this signifies a red flag. This means the lending agent does not know what he or she is talking about, or does and is hiding critical facts from you, where you could end up paying for that lack of knowledge. You need to know these things too, how to perform the calculations, and assess the economic alternatives.

Nothing is cheap in the worlds of home and car maintenance. Here you'll be dealing with contractors and mechanics over the costs of repairs and materials, which may lead to overbilling. Do-it-yourself is more than about satisfaction of accomplishment and the therapeutic joy of working with your hands. It is also about saving boatloads of money, financial necessity, technical insight, and personal responsibility.

Lotteries and casinos are good for the state or other entity in which they operate but not for you. They benefit from the laws of probability and human weakness. The odds are stacked against you from winning by design. Of course, they don't want you to know this and assume, correctly so, that most people are too mathematically deficient to understand their handicap. You'll notice as you get older that people who habitually play the lotteries and/or frequent casinos are the kind of people who should not be doing so.

All scams are based on trust. They apply to both your personal life and your future professional life. Scams are most effective when people are desperate, unknowledgeable, and/ or fail to do their due diligence, particularly in doing their math. One type of scam comes in the form of pyramid or "Ponzi schemes", which is a type of proliferating investment where one victim is required to recruit additional victims (although they don't know they're victims until it's too late) to receive the so-called returns or benefits. I know a few people who have fallen for these tricks and were burned because they failed to do their math.

> ***There's a sucker born every minute.***
> - Often misattributed to PT Barnum, American showman, 1810-1891; most likely uttered by banker and horse trader David Hannum, 1823-1891, who was associated with the famous *Cardiff Giant* hoax of 1868-1869 (an interesting research topic for you)

Bernie Madoff, a name that you probably don't recognize, was a notorious investment huckster who convinced investors—many of them prominent luminaries—to put their money in his fraudulent investment company's funds that "earned" consistently high "returns". Being convinced by Bernie was easy since he spoke with extreme, articulate calm and confidence, and the numbers displayed were so reassuring. In this case, according to the whistleblower who presented the case to the US Securities and Exchange Commission (SEC), it wasn't so much that the investors were duped, but that the SEC was itself. *"It was all made up and his story was so fanciful and far-fetched that the SEC should have seen through it immediately. And they didn't."* and *"The SEC is staffed by lawyers who don't understand the mathematically complex financial products that are traded on the markets these days."[3]* (Bernie was subsequently convicted of money laundering, securities fraud, and other charges, and was sentenced to 150 years in prison where he later died in 2021.)

> *Finance is mostly smoke and mirrors but Math is truth. When presented with a potential financial or accounting fraud, the first thing I do is figure out what math tests I need to use to test whether it's a legitimate investment or fraudulent. The math never lies.*
>
> - Harry Markopolos

"Financial products" are another source of being taken advantage of if you don't scrutinize them thoroughly and mathematically. There are plenty of these "sophisticated investments" or "alternative financing options" that you probably don't need, now or later in life. Some relatively simple types are reverse mortgages, and adjustable-rate mortgages. Other nonessential kinds include debt management plans, rental car insurance (since your own car insurance may already cover rentals), identity fraud protection, predatory loans, ...the list goes on. Some of these things may be completely useless to everyone, whereas others could be

---

3   Harry Markopolos, 1956-present. "Madoff Whistleblower: SEC Failed To Do The Math." Morning Edition transcript, National Public Radio interview, March 2, 2010, https://www.npr.org/2010/03/02/124208012/madoff-whistleblower-sec-failed-to-do-the-math. Harry is a certified fraud examiner and a forensic accounting and financial mathematics consultant.

appropriate for some but not all. You need to understand the math behind them to make sense of them. Yet they could be presented to you by reassuring and cool "experts", or famous celebrities that you "trust", via professional marketing formats, where there is a conflict of interest, and the peddlers know that many people are deficient in their math skills and general judgment and tend to make emotional decisions rather than logical ones.

> *Assume fraud first, until genius can be proven by running the numbers and seeing if they add up. With fraud schemes, the numbers never add up, never add across, and never pass a basic sniff test.*
> - Harry Markopolos

Let's talk a little bit about lost opportunities. Have you ever wondered how rich people became so wealthy in the first place? Granted, some inherited their wealth or were "born on third base" with lots of academic and financial support at home when they were growing up. Many aren't as rich as they seem but live beyond their means. Others simply worked hard, owned businesses, and/or made a fortune by making intelligent investments of various sorts—with both their money and *time.* Some people have made investments and did not become rich, not so much due to bad luck, but because their investments were not so intelligent. They took unreasonable risks based on flawed judgment and lost (been there, done that, but fortunately on a small scale). The intelligent investments were founded on knowledge and calculated risks where the investors studied the markets, the working of various financial mechanisms, economics, investment choices, companies, real estate, trends, and so on, and then made logical, rational, disciplined data-driven decisions using mathematics, among other things. They did not make their investment decisions based on emotions, following the crowd, what other unqualified people told them to do, gut feelings, anxiety, or panic. In case you ever ask, *"Why didn't someone ever tell me about these things?"* or *"Why didn't they cover investing and personal finance when I was in school?"* Personal finance and investing are not typically taught in most high school curricula. And yet, mathematics was taught in school, which is a critical component to these. Learning how to study math, as well as other subjects—inside and outside of the classroom—is another thing not commonly taught in schools. This is something that I will tackle in the next chapter.

Public Finance and Government. Government officials at the local, state, and federal levels are not immune from math-related mistakes, mismanagement, scams, and corruption. Hence, you as a taxpayer (or future taxpayer, whether you like it or not) and beneficiary of government services, or as a government employee or military service member can be taken advantage of by them, directly or indirectly, deliberately or unintentionally, and they can be taken advantage of as well—where you may end up paying the price.

The ignorance of some politicians and public officials, and/or their lack of research requiring math can lead to harmful decisions where they thought they were doing the right, prudent thing, but disregarded, or weren't aware of, red-flag warnings related to public finance. The results can lead to: massive debt; higher, unnecessary taxes to pay off the debt; and bankruptcy. Corruption, extortion, bribery, fraud, are criminal matters that, needless to say, can negatively impact you as well. Rarely does a day pass where one can't find examples of this mischief in the news uncovered by government employees, reporters, whistleblowers, investigators, company insiders, and private citizens.

Bowing to pressure from lobbyists and special-interest groups to achieve short-term political gains with long-term consequences helps drive poor decision-making inside the government where you as a citizen and consumer can incur the penalties.

Criticizing public officials and those in management positions in organizations and in the private sector is, of course, very easy. Are the reporters who make these cases correct? Are the courts that hand down the decisions being fair and impartial? Would you have done anything differently if you were president, a congressman or congresswoman, judge, alderman, selectman, mayor, or other official in their situation? To answer these questions and make responsible conclusions, you need a basic, well-rounded education and a lot of life experiences, where a key part of your education includes a sufficient degree of math. Without the math, you will not be able to understand and make sensible, conscientious conclusions on the decisions made that entail basic science, accounting, and finance. You will be dependent on "experts".

Your job is to pay attention and challenge these officials directly in the courts or at the voting booth. Democracy is messy; it's *sausage making* as they say. But if you're the victim of poor public financial arrangements or of corruption, accountability and remedies are in order, especially for the latter. Thankfully, we have the free press, law enforcement, astute individuals, and a number of watchdog organizations to help monitor and manage things, albeit imperfectly. Some cases in point, at least at the federal government level, entail the Federal Reserve Board, SEC, Office of the Inspector General, and the Food and Drug Administration.

Health and Well-being. We all need clean air, clean water, safe food, shelter, clothing, and security to survive. We all want good health, since achieving happiness without good health becomes more difficult. Unfortunately, people in this world offering goods, services, and ideas to you, which affect your needs, can take your health away if you're not careful and sufficiently knowledgeable. An adequate level of mathematical proficiency along with math-dependent science, among other experiences and knowledge, can only help you preserve your health and well-being. The actions (or inactions) of government agencies (for example, the Department of Agriculture, Environmental Protection Agency at the federal level) can work for you or against you in this regard.

Water is a key ingredient to all life on Earth. Without it, we humans die, never mind other animals and plants. But for water to be beneficial to us, it must be clean; that is, potable. You as a consumer in the US cannot take clean water for granted. Recent history is filled with examples of how contaminated water caused severe health problems in people. They include: the water crisis in Flint, Michigan; the coal ash spill in Kingston, Tennessee; the Love Canal in New York; and aging lead pipes in many US municipalities. The story will continue.

Food, of course, is another factor that keeps us alive, but like, water, it must be unspoiled, nutritious, and uncontaminated for it to do us any good. We need to understand the world of agriculture, fisheries, processing facilities, and the like, as well as the agencies set up to oversee this wide-swathing industry. There are many facets to be aware of in this regard since there are so many factors that can taint our food chain that are likely to only get worse, from microplastics, to mercury, to heavy metals, to pesticides, and a host of other contaminants.

An additional basic need for us is shelter. Math can help us evaluate shoddy construction (at least to a minimal extent), analyze financing decisions (mortgages, rent vs. owning, other smaller loans), fire hazards, general safety issues (electrical, gas, plumbing, toxic materials), and in dealing with contractors, their subcontractors, building inspectors, real estate agents, property managers, and suchlike housing-related intermediaries. Math can also help us understand building codes, technical drawings, as well as legal restrictions, and limitations on housing issues.

Water, ground, and air pollution directly affect us all. Too many of us breathe bad air and live near contaminated ground areas and waterways that can have a harmful effect on our health. Contributing to this mess may be large-scale engineering projects with unintended consequences such as from mining waste, oil spills, environmental impacts from dams, and being subjected to the harmful aftereffects of nuclear power plant disasters.

Our health care system is overly complex and expensive; it's due for an overhaul. Of course, this has been said for many years. Fixing it is easier said than done and has been a big topic in every presidential debate that I can remember. If you suffer a major illness or calamity, and are not aware of the pitfalls, you can easily go bankrupt, and be subjected to wrong diagnoses, procedures, and prescriptions. Insurance and hospital billing procedures can be opaque, capricious, and unfair. And providers, in general, can take advantage of you—deliberately or inadvertently. You should have a basic understanding of what they're prescribing (or cutting!) and why they're prescribing it. An inherent conflict of interest exists between patients and providers. Things could be prescribed to you that you don't need. In fact, due to the conflict of interest, doctors can be afraid to tell patients what they really need (which, in many cases, is a kick in the behind!). Doctors and nurses are human, and they make mistakes. They're subjected to stress, fatigue, emotions, and frustration like all of us, and can get burned out from long work hours, threats from lawsuits, never mind administrative and procedural minutiae.

Business-to-Business Transactions. Risks and hazards are everywhere in business-to-business transactions. If you're going to be a business owner, you have to always be on your guard. If you're not vigilant, you could be subjected to: employee graft, incompetence, and indolence; underperforming sales representatives; contractual legalese that, if not read carefully, can get you into big trouble; getting "stiffed" (not getting paid) by customers; the negative effects of poor financial management and documentation; being at the losing end of negotiations; regulatory penalties at the local, state, and federal levels; to name some. You will want your business to thrive, prevent bankruptcy and lawsuits, and avoid laying off good people. In the business world of multitasking, making mistakes is easy. You will always be imperiled by pressures and constraints from multiple sources that you may have no control over.

Additional Reasons and Considerations (and Harping)

Let's face it, you're going to die someday; we all are. But you don't want to be taken advantage of or be unhealthy or sick while you are alive. Instead of being at the mercy of others or placing the blame on them for those things that you have some degree of influence over, adopt the mindset of self-help, self-reliance, and personal responsibility for avoidable problems. On the other hand, you don't want others to cause problems for you and negatively affect your personal freedom, finances, health, and well-being leaving you vulnerable. Math will help you prevent being taken advantage of by overconfident-yet-eloquent news pundits, politicians, coworkers, lawyers, mortgage brokers, financial advisers, health care professionals, contractors, friends, family members, and luminaries who do not know what they are talking about, or do and are trying to pull the wool over your eyes.

You will never have total autonomy over your life. There will always be problems, and more are coming. Paying attention, listening, reading, watching, questioning, challenging, and doing your homework—all, of which, requires math (among other things), including a certain degree of advanced math—won't guarantee happiness, but will minimize problems and being caught off guard. On challenging and questioning others, be careful on how you go about this. Your own beliefs and conclusions may be right or wrong. You also need to know what you are talking about, and there's a way to disagree without being disagreeable. Proceed humbly.

The three stated reasons to study math and take it seriously are necessary but are probably inadequate for you. Let me supplement them with some additional reasons, harping, and words of wisdom that more than likely no one else in your young life has ever conveyed to you, or ever will. Remember this: *"They that won't be counseled, can't be helped"*, as Poor Richard says, and further, *"If you will not hear reason, she'll surely rap your knuckles."*

I Don't Like Math or Science! Here are some common sentiments I hear frequently:

- *But I don't want to be an engineer, scientist, doctor, finance person, mathematician, or any other person who uses math occupationally!*
- *I don't like these subjects, and I don't want to deal with calculations, complexity, frustration, and problem solving!*

- *I just don't like math!*

And? Who says that you have to like math to do it, or anything for that matter? Many students (and adults!) don't like eating their vegetables, exercising, doing housework, or attending to their personal hygiene. Do them anyway. Life is tough and was never meant to be easy. For you and your generation, it will likely get a lot tougher. Let's destroy some myths and misconceptions right now and shed some light on reality.

- We learn many times by failure, not through success (although, admittedly, the latter is more enjoyable)
- *Work* is not the same as *fun*, but sometimes it is and can be therapeutic; if work was always fun, people would pay $3,000 to go to work instead of Jamaica for vacation
- *Need* and *want* are not always the same thing
- There's virtue in drudgery, and when the work and drudgery are over, and we've met our goal, we become happy

By studying math diligently and taking it seriously, you will not die; your brain will not explode. In fact, your brain will benefit many times from exhaustive work. What heals it is sleep, proper diet, exercise, and recreation. Diligence towards math will provide you with an investment return in many direct and obvious ways, as well as innumerable indirect and not-so-evident respects. You will only receive positive rewards in the end.

Can You Still Navigate? Can you still navigate in the world without a solid foundation in mathematics, limited to arithmetic and some introductory algebra? Of course, and many do quite well for themselves. They come a dime a dozen. Just be aware that their method is not guaranteed to work for you.

But why muscle through life from a position of weakness and ignorance? Why be at a disadvantage? *Ignorance is bliss* is true—until it isn't. It's a fallacy. If you don't get burned by your ignorance (and hopefully not by your apathy), someone else will, directly or indirectly, right away or over time, which could include your employer, your employees, coworkers, children, spouse, and society at large. This is true in the military, business, medicine, in your home, and, well, everywhere. Let's make an analogy. You don't need an MBA to run a business, but the knowledge it provides can only help if you want your business to thrive and grow. Knowledge can only help; it will never hurt.

If you get the basics down when you are young when the opportunity exists, everything is gained, nothing is lost, and you have firm groundwork to build on. Other things can and will wash away, but not your foundation; that is, your core education. You may never get the opportunity to formally study math again. After high school, you may find yourself in a situation where you cannot or will not take up the opportunity if it does arise due to lack of time, money, and the required effort. Life and responsibility tend to get in the way. Your education up until now has likely been paid for by other people's taxes. After high school, the dynamics change.

You and/or your family will have to pay. And if your post-high school education requires math, either because your intended occupation demands it, or your school, college, or university mandates it, know that remedial math is costly—monetarily and emotionally.

The Proverbial "Successful" Person. I recall an English professor being interviewed on the radio some years ago, although I forget the particular show or topic. He was probably a successful author of some sort and was self-assured, convivial, eloquent, relaxed, and said something that I found to be both profound and disturbing. Paraphrasing, he said, "*I never truly needed math beyond basic arithmetic, and of the math I had, I can honestly say that I rarely, if ever, used it or needed it.*" Really? I hear a lot of this, even among some engineers since, admittedly, much of what a typical engineer does on a day-to-day basis is not highly technical, per se (even though many wish it was). The English professor committed a number of all-too-common sins, including overconfidence and complacency. His view was normal and can be forgiven, but this doesn't mean that you should subscribe to his way of thinking. Too much is being taken for granted with this attitude, which is easy to do for what we all use routinely or subconsciously.

The English professor succumbed to the *I'm happy, and there's nothing wrong with me* mentality. There's something wrong with all of us. If you don't believe this, ask people who don't like you what they think of you. As Poor Richard said, "*Love your enemies, for they tell you your faults.*" We should always be looking for self-improvement opportunities over our entire lives. Otherwise, we become self-delusional, overconfident, complacent, and, thereby, vulnerable. As the adage goes, *If you're not getting better, you're getting worse.* I would argue that the English professor is educated, but in a classic, unbalanced way, and that he has not lived to his potential.

The English professor's statement begs several questions, such as:

- Is this the same guy who advocates for memorizing Shakespeare, vocabulary, and poetry that you may never hear of, see, or use again? (By the way, I'm not arguing against the value of these things. I'm just trying to make a point.)
- Is this the same guy who travels internationally without considering how he got there? (Think of the aircraft, engines, airline and airport operations, logistics, fuel system, infrastructure, navigation systems, air traffic control system, meteorology, radar, and so forth, none of which are possible without mathematics.)
- Does he write using a computer?
- Can he make sense of complicated, math-dependent scientific and financial matters?
- Is his eloquence, self-assurance, and persuasiveness adversely influencing administrators, public officials, and the citizenry at large—technically or financially?
- Could he be subjected to being taken advantage of in a subtle, indirect way—intentionally or unintentionally—as described before?

Here's another anecdote that should prove illustrative. I recall a discussion I had with an engineering technician at a company I worked at who was describing his encounter with his child's history teacher. He had told the history teacher, who was teaching ancient history at the time, that he wanted his child to be taught only recent and "relevant" history because ancient history was—in his mind—irrelevant. I was taken aback. The engineering technician was essentially saying that he knew more than the history teacher about history and the education system, and that studying the origins of western civilization through that of the ancient Greeks or Romans was a waste of time. I couldn't agree less. We can't understand the present unless we understand the past. Understanding the 2000s can't be done without understanding what happened in the 1960s, 1940s... all the way back to the cradle of civilization (and even before that), to put things in context, make sense of the world, not repeat mistakes, and make decisions accordingly. Think of a doctor. He may not need to use or remember the details of molecular biology, organic chemistry, pharmacology on a daily basis, or what arrector pili muscles are (think "goose bumps"), but his extensive premedical and medical school training gives him insight, understanding, procedural know-how, and options to investigate problems further to treat his patients properly.

Math is the same way. Like history, even if you don't like math or use it every day, including the *advanced* kind, you need it to understand the world we live in and form *correct* opinions accordingly. This applies to the movers and shakers of this world, namely, our representatives in government, business leaders, military leaders, and the like. Their decisions—flawed or not—affect us all for good or ill.

If you define a "successful" person by their accumulation of worldly wealth, influence, and power, know that some who have acquired that wealth did so by luck or by being unscrupulous. (I would argue that one can be dirt poor and be a success, depending on how one lives his or her life. I consider Mother Teresa[4] to have been a success.) Ambition and authority positions do not equate to competence. In cases of obvious incompetence, one always has to question how these people accumulated their wealth and position of authority. Some of these people are simply astute at knowing how the financial, business, and/or political world operates such that they take advantage of that knowledge. Many people with wealth and authority are, of course, simply hardworking, competent, talented, driven people who sacrificed a lot to achieve their status. What's my point? If financial success is your goal, be careful, move with caution, work hard, and think deeply about the type of person you want to become, know the value of math, and don't become "successful" for its own sake. Let's leave this with some of Poor Richard's sagacity:

- *"Many, without labor, would live by their wits only, but they break for want of stock; whereas industry gives comfort, and plenty, and respect. Fly pleasures, and they will follow you. The diligent spinner has a large shift; and*

---

4   Mother Teresa, aka Saint Teresa of Calcutta, founder of the Order of the Missionaries of Charity serving the destitute of India, 1910-1997

now I have a sheep and a cow, everybody bids me good morrow."

Conclusion

The three main reasons I provided to study math are all interrelated. If national aims are met, citizens benefit. We will never eliminate all problems, but we can certainly minimize them through positive and negative incentives. Our government works for a reason, although not always flawlessly. If legitimate selfish aims are met, the country benefits. With a sufficient level of math, you will be able to survive and, hopefully, prosper. Chances that anyone will be able to take advantage of you will be reduced, even if you don't want to be someone who uses math regularly in their work. Episodes for malfeasance, incompetence, and injustice will decrease. Knowledge, including mathematical knowledge, never hurts and can only help. It's a competitive discriminator.

You only have limited opportunities for school. Your years on this planet are finite. You'll find that time seems to go much faster as you get older in a way that you wished it wouldn't. Don't screw it up! Don't squander this structured opportunity that we call *school*. Colleges have started to crack down on granting college credits to remedial math courses that should have been mastered in high school. (For the record, my remedial math courses that I mentioned in the *Introduction* chapter did not factor into my engineering degrees or MBA, and they shouldn't have. But they did cost time and money!)

I'll continue to harp: Math is necessary, yet is insufficient on its own, but it supports just about everything in both obvious and nonobvious ways. It's a language and a set of tools.

What industry, government, the military, customers, stockholders, the public (from their public servants and others who affect their lives), and academia want—and what we want from ourselves—are people who:

- have integrity
- can solve problems sensibly
- can get along with other people
- applied themselves throughout life

- are qualified and experienced, or are learning from their experience with a purpose

- have good judgment
- can think for themselves
- add value to the organization
- have a strong work ethic and want to be in their roles
- have the right attitude, paid attention in school, did their homework while the opportunity existed, are self-motivated, and self-managed

On the matter of being able to solve problems—those things that life is full of—we all need to be able to:

- think and reason critically, logically, analytically, objectively, and abstractly
- understand and use arithmetic, magnitudes, shapes, and logic
- demand requirement of proof
- make judicious, objective decisions
- work from an advantage of knowledge, not ignorance
- have a foundation from which to learn further, add value to our family, employer, employees, government, military, fellow citizens, country, the world at large, and ourselves

Showing unacceptable ignorance where you should know, as in requisite math knowledge and skills, will render you ineffective to an incredulous audience (another area where I can speak masterfully), and can cost you in lost employment, business, and in so many other ways.

> **You're in school to get educated, not to become a trained animal.**

Pay attention to the news and use multiple, *reputable* news sources. Seek the truth and not self-affirmation. Knowledge equals power and reasonable self-confidence. Life and society can be messy. (Ask an honest politician who is trying to do some good.) Our political system is based on compromise; it can't function without it. Math helps facilitate this messiness.

For society at large, wishing a problem, such as a safety concern, will go away can be self-defeating and downright dangerous. Very recent history with nuclear facility catastrophes, bridge and building collapses, oil rig fires and leaks, and financial crises all speak to this point. You have a host of long-term safety and survival challenges awaiting you dealing with climate change, pollution (air, water, and soil), potable water scarcity, pandemics, and biodiversity reduction, which are all exacerbated by population growth.

If you start thinking differently about math, which I hope you do as a result of this book, by considering the history, definitions, purpose, and yes, beauty of mathematics, it will become less painful and, expectantly, more enjoyable.

I hope that I've answered the questions you may have had about the reasons to study math, why it's important, and to take it seriously. If you feel that this chapter is still incomplete, perhaps the next one will help fill in some of the gaps.

5

How to Study and Learn Math

When I hear, I forget
When I see, I remember
When I do, I know[1]

How to Study and Learn Math is the most important chapter in this book. Your studying and learning techniques will make or break you. I never met a high school student who had trouble in math—with no troubles in any other subject—who studied math correctly. The remedy may not be to work harder, *per se,* but to work smarter. Books, web pages, companies, and videos abound on math-specific study habits. What differentiates this book from others is that I'm conveying to you what you need to hear in a way that no one else will ever likely do, and what I wish someone had told me when I was a junior high and high school student. Other books and resources will not communicate the same message since most of it you may not want to hear; they don't want to risk alienating you. Your study habits more than likely need improvement anyway.

---

1   The translation and author of this famous quote are debatable, but the author is probably Confucian philosopher Xunzi, circa 300-230 BCE

I acknowledge that we're not all the same and that we learn differently; one size does not fit all. No argument. In the process of improving, you will learn your strengths and weaknesses. Regardless, you must do math work in a way that is different than for other subjects[2]. Avoid hearing what you want to hear.

How should you study and learn math? In a nutshell:

1. Get organized and manage your time
2. Read the book
3. Pay attention, take notes, and ask questions
4. Do the homework
5. Prepare for exams properly
6. Change your attitude and approach

You already knew this? Read on to get the details from a different perspective.

Get Organized and Manage Your Time

Order. Organizing, scheduling, and time management are key to mastering your subjects, including math. Reject the all-too-common bad habit of shifting from one subject to another in a haphazard and casual manner, and then cramming for exams at the last minute. This is a recipe for a letdown. Serendipity has its place in life but not here. Rather, adopt an ethos of self-regimentation and discipline. Doing

---

2 Arguably, a portion of this chapter can and should apply to all subjects. But significant content that is unique to math and math-dependent technical subjects is presented in this book dedicated to math.

so can be hard but not impossible, and it gets easier with structured and persistent implementation, reinforced with successes along the way. Learn to prioritize. That is to say, determine what is important, not important, unessential, and what must be done immediately. *Do what is due first*, not necessarily what you want to do. Find a suitable study place, remove or minimize distractions, organize your class materials and physical space, and schedule your time so that you can obtain your ultimate goal. These small but significant steps will yield a huge payoff in the end. Remember that the ultimate goal is to get done what needs to get done, and learn what needs to be learned, yet still have time for recreation, sports, extracurricular activities, proper nutrition, adequate sleep, friends, and family. You need to invoke the F-word: *Focus*.

Location and Class Materials. Find a clean, quiet study area with adequate lighting be it at home, school, library, or wherever works. Organize your study area since organization promotes efficiency. Keep in mind that you need at least three physical things in a particular subject:

1. Textbook
2. Notebook
3. Folder for papers

Notice that I didn't mention a computer, or any other electronic device. The omission is intentional. Use a computer only if mandated by your math class for homework.

I'll have more on your textbook and notebook in a little while. Now regarding your folder for papers, your papers should be in some degree of order—numerically, chronologically, and categorically. You should include a syllabus (more on that later as well) and have your papers follow the syllabus somewhat. If your teacher is disorganized, do your best to compensate. You should have sections for graded quizzes and exams, worksheets, and homework papers. Simplify and consolidate. Perform a periodic purging to get rid of irrelevant papers and excess. You should have different but easily identifiable folders for each of your subjects, as in different colors or specific markings such that if you had to grab a folder from a pile when you're rushing out the door to catch the bus, you can easily take the correct one. If you must use a computer for schoolwork, your file management system should be equally organized and consistently maintained and evaluated for purging and updating.

Once you have your study location and materials situated, your desk or table needs to be organized. Remove the clutter and excess. Adjust the room's temperature and lighting as much as possible. If there is no air conditioning, get a fan. If it's too cold, put on a sweater. If there's too much noise, close the door or window, or find a better place.

Your state of organization reflects your priorities, organizational skills, and personality. A scattered work area and disorganized materials reflect a scattered brain, which is has a negative impact on studying and learning math properly.

Remove or Minimize Distractions. Distractions work against study habit improvements, deep thought, and learning. You need to remove those that you have 100% control over such as:

- Earplugs (music)
- Computer (when not required as part of your homework), tablet, and/or cell phone
- TV
- Pet

Learn the benefits and peacefulness that comes with silence. Don't be like "everybody else" and succumb to weakness. Adopt the mindset of those schools and parents that ban the use of electronic gadgets. Understand their harmful effects on the developing brain. (And yes, your brain is still developing, and will be for some time.) You will learn and retain so much more without these distractions, the goofing off (music, texting, social media, videos, etc.), and the communicative disruptions that they bring about. I understand that many teachers impose computer-related homework to their students. In these cases, you should just do the assigned work and shut the devices off immediately afterwards. If your parents or guardians do the opposite of what I'm prescribing, don't follow their example. One way to counteract distractions, in general, is to schedule your time.

Schedule Your Time. You can't control everything so you must adjust and focus. Do you have a study period during the school day? Then study. Take advantage of your time of relative freedom to schedule your activities throughout the week. You may never have enough time again. Later in life, you could find yourself at the mercy of demands for your time that you have no leeway over whatsoever.

Instead of worrying about studying and homework, especially the kind where there's uncertainty and you have to "figure things out" as with math, schedule this homework like all other types but account for the frustration that can come with it; that is, account for the "figuring out" and frustration time (for example, add an extra 30-60 minutes up front). In this way, you'll have time for other things in life that are also important.

After my tour of duty ended in July 1988 at the US Embassy in Paris, I flew back from Europe headed to the headquarters for embassy Marines in Quantico where I was to administratively finish up my second four-year enlistment in the Marine Corps prior to my discharge from active duty. I was to begin my formal freshman year at UMass Boston about a month later. During this flight, I was fortunate enough to sit next to a lawyer from Texas who was returning from his vacation in Austria. I asked him about law school and his study habits while a law student, where one of his lessons learned sticks with me to this day. He explained that he did a lot of goofing around as a prelaw undergraduate, but he became a much more focused student in law school where he was preparing for his lifelong career. He said (paraphrasing), "*You can get a lot of homework and studying done in a planned, scheduled, laser-focused four hours, versus claiming to have studied for four hours, when in fact, you talked with your friends, looked around, daydreamed, dawdled, walked about, and so forth.*" His observation should not have been a revelation to me—or you—yet we all know it to be true.

Another anecdote centers around my own lessons learned while an undergraduate at UMass Boston. My study habits at the time were not bad but needed improvement. On weekends, I would do all my chores, workout, run errands, perhaps play basketball with friends in the mornings and early afternoon, and put my homework off until the afternoons and evenings. In other words, my anxiety and worrying about homework (think lab reports, exam preparation, papers, computer programs, difficult engineering problems) followed me all morning and into the early afternoon until I started. Then, when I wanted to enjoy the somewhat warmer weather, quieter atmosphere, and late-day or evening entertainment options of various sorts, I had to hit the books and deal with difficult problems while exhausted. This was all exacerbated by the all-too-frequent realization that the time I allotted for my academic ordeal was insufficient; the problems I encountered required more of my time than both the day and night allowed. I don't even want to mention income-earning work demands or an entire weekend dedicated to the Marine Corps Reserve. I recall discussing my time-starved dilemma with a fellow engineering student who told me of her study routine, which consisted of starting her homework first thing in the morning when her brain was operating in a state of optimum alertness. Once finished, say by noon or 1:00 pm, the rest of the day was hers. The light bulb went on and I adopted her method for the rest of my undergraduate and graduate career with great success. If my homework was not completed in those four to five morning or early afternoon hours, I would certainly have to shift other obligations around, but at least my academic concerns and anxiousness—which translated directly into Grade Point Average, money, post-graduation employment—were mostly alleviated. My work and Marine Corps Reserve obligations had to be met, and yes, things became very complicated and stressful during those times, but I adjusted to make things work in challenging ways. (I never want to go through that again. Thank goodness I was still relatively young.) Adopt this approach of doing your homework in between classes, and on weekend mornings, if possible and you will reap the rewards. (I'm aware of sports practice, dance rehearsals, and other commitments.) I'm sorry to say but the Saturday- and Sunday-morning cartoons will have to wait. Doing your homework and cramming for exams late into the night is typically unnecessary.

> *In studies, whatsoever a man commandeth [sic] upon himself, let him set hours for it.*
> - Sir Francis Bacon

To say to yourself that you don't have the time to do this or that important thing is probably not true. Ask an unbiased, outside critic—or better yet, an enemy—to evaluate your daily routine and habits to determine where opportunities exist for improvement. You'll be surprised at what you hear, unpleasant as it may be. I recall working with one high school freshman some years back on her writing assignment. The assignment was not very lengthy, but I was trying to explain to her the benefits of writing an outline for the task first to approach the actual writing in a methodical, disciplined, logical, and thoughtful manner. She did what most people do and just impulsively started writing, where the result was not bad but was somewhat rambling and disorganized. Her response to me at the outline suggestion was, "*I don't have time for that.*" A 10-to-15-minute investment into a brief outline could have resulted in a comprehensive, well-written, organized paper that would have yielded her praise and an *A* grade from her teacher. When a student says, "*I don't have time for that*", this typically means one of two things:

1. She truly doesn't have time because she squandered earlier opportunities, or
2. She does have time but is making up excuses to avoid necessary, unpleasant, aggravating drudgery, and is succumbing to weakness and laziness

Did she have time for texting, social media, clowning around, and watching TV and movies?

You probably don't have much experience with painting a room or a house, but the actual painting is the easy and desirable part. The coveted, final product is obvious and immediate. Preparation is the more important, time-consuming, difficult, and unappealing part. When proper preparation is not done (sanding, patching, spackling, crack treatment, scraping, smoothing, deglossing, cleaning, and priming) the paint job looks like an unprofessional disaster or will become one later. Scheduling your week is like constructing the outline for a writing assignment and the preparation for painting. Generating the schedule and sticking to it can be difficult but necessary and produces great results—immediate ones and those in the future. Skip these steps and you end up with mediocrity. Figure 1 shows an example schedule of a high school student who is fruitfully engaged in academics, sports, family, friends, and still holding a part-time, eight-hour-per-week job.

Time	Monday	Tuesday	Wednesday	Thursday	Friday	Saturday	Sunday
6:00 AM	Breakfast, Prep, & Transport	Breakfast, Prep, & Transport	Breakfast, Prep, & Transport	Breakfast, Prep, & Transport	Breakfast, Prep, & Transport	Sleep	Sleep
7:00 AM	Homeroom	Homeroom	Homeroom	Homeroom	Homeroom	Breakfast	Breakfast
8:00 AM	US History	US History	US History	US History	US History		
9:00 AM	Pre-calculus	Pre-calculus	Pre-calculus	Pre-calculus	Pre-calculus	Homework	Church and Sunday School
10:00 AM	Spanish	Spanish	Spanish	Spanish	Spanish	Chores	
11:00 AM	Lunch	Lunch	Lunch	Lunch	Lunch		
12:00 PM	English	English	English	English	English	Lunch	
1:00 PM	Chemistry	Chemistry	Chemistry	Chemistry	Chemistry		
2:00 PM	Study Period	Study Period	Study Period	Study Period	Study Period	Job	Recreation, Friends, Family, Lunch, Dinner, and Miscellaneous
3:00 PM	Basketball Team	Basketball Team	Basketball Team	Basketball Team	Basketball Team		
4:00 PM							
5:00 PM	Transport	Transport	Transport	Transport			
5:30 PM	Recreation	Recreation	Recreation	Recreation			
6:00 PM	Recreation	Recreation	Recreation	Recreation	Job		
7:00 PM	Dinner	Dinner	Dinner	Dinner			
8:00 PM	Homework	Homework	Homework	Homework	Recreation, Friends, Family, Dinner, and Miscellaneous	Recreation, Friends, Family, Dinner, and Miscellaneous	Homework
9:00 PM							
10:00 PM	Recreation	Recreation	Recreation	Recreation			Recreation
11:00 PM							
12:00 AM							
1:00 AM							
2:00 AM	Sleep	Sleep	Sleep	Sleep	Sleep	Sleep	Sleep
3:00 AM							
4:00 AM							
5:00 AM							

Figure 1 – Example Schedule[3]

---

3   I'm aware that you may have certain classes every other day or maybe once or twice a week, but I think you get the gist.

If all you do is work, you risk becoming a dullard. If all you do is play and lollygag, you're liable to become a know-nothing, unhappy, boring, parasitic deadbeat, who is a dullard of another sort. Life is about balance. Unfortunately, *balance* is up for interpretation and can be misapplied. Recreation is necessary for mental health and is beneficial to learning, social interaction, creativity, and success in academia and in the work world. Stress, anxiety, deadlines, and frustration can also be good for your mental health as well provided they're not constant. Play is extremely important in your younger, formative years, but, as I'm sure you're beginning to figure out, much less so as you get older. Sports, household chores, part-time jobs, friends, recreation, sleep, and family obligations are important and need to be scheduled in. How? Get your priorities straightened out. Before that, you have to figure out what your priorities are, and then distinguish between the important and unimportant, which isn't always easy to do.

You also need to schedule in a physically healthy lifestyle, which includes proper nutrition and exercise. The US Department of Agriculture's *MyPlate* (www.myplate.gov) is a good place to start to define proper nutrition. Avoid snacking. Shoot for a multicolored plate of food at three balanced meals per day with satiety value that will counteract the desire for snacking. Always keep in mind, the more you eat of the good stuff, the less you will want of the bad stuff. Also, eat junk and your body—including your still-developing brain—will perform like junk. Your appearance and performance will reflect your diet.

Good books and information on the physical and mental benefits of exercise are abundant. Our bodies were not designed to be sitting around looking at computer screens all day. Your body needs to move—a lot. Walking one mile per day is not only inadequate but unbalanced. Your physical activity routine should aim for skeletal muscle, aerobic, and anaerobic exercises, joined with flexibility exercises, healthy competition, among other things, but also considering injury prevention. (A good sports program offers all these things.) Do you lack energy? Always remember the human energy paradox that no one likes to hear: *You have to expend energy to gain energy.*

Recreation, nutrition, and exercise are great, but proper sleep also matters. Sleep does wonders after experiencing an exhaustive day at school followed by a night of frustrating homework. Never mind other health benefits and necessities, sleep also organizes thoughts. The "delayed effect" and cogitation are part of the learning process. Study, think hard, wrestle with a problem, research, dig for information, get extremely frustrated and anxious because you can't figure something out, especially over a long period of time, throw up your hands, and then go to sleep. When you wake up, your problem may or may not be solved, but you certainly learned a lot. And when the problem is solved, maybe right away or the next day in class (or even later than that), you get to experience that *Eureka!* moment, which is extremely gratifying.

The list goes on. The point of this whole passage is that you need to schedule and manage your time—all of it—not only for your academics but also for those things to maintain and improve your physical and mental health, and to earn an income. Without taking care of yourself properly, your schoolwork will suffer. Likewise, only benefits will be realized with the right amount of exercise, recreation, nutrition, and sleep.

In summary, schedule your time and stick to it so that you not only earn good grades, but still have time for those other things that admittedly tend to more enjoyable.

Read the Book (and Syllabus)

Textbooks. I realize that many schools don't issue math textbooks as they used to, if at all, due to their high cost. Hence, math teachers hand out worksheets or refer students to computer-based resources instead, which is not wrong but not ideal. However, if you have been issued a textbook, chances are it's a good one. If your school didn't issue you a textbook, get one. Ask for one. Buy one. Borrow one. Or check one (or a few) out from the local or school library. You need one from a reputable academic publisher (for example, McGraw-Hill, Houghton-Mifflin, Wadsworth, Addison Wesley). Listening to the teacher, taking notes, working from assigned worksheets are necessary but insufficient for comprehensive learning and studying.

The Syllabus. You don't have a syllabus? Get one of those from your teacher. Ask politely, tactfully, and state your reason, because (let's hope this is not the case) he or she may not have one. Teachers typically respond well to students who want to work hard and learn. A syllabus will allow you to anticipate what is coming, how the course is organized, what to cover when reading ahead, and your teacher's expectations. The textbook and syllabus will allow you to synthesize your coursework, complement the teacher's instructions, and prepare for lessons and exams.

Read the Textbook—*all of it*. Now that you have a textbook, read it, yes all of it—methodically and incrementally—including the introduction, footnotes, sidebars, appendices, and summaries. Consider this as part of your homework—assigned or not. Refer to the textbook often and browse it regularly. Someone (namely, the author) with experience and wisdom in math is trying to tell you something. A reputable textbook is much better than what is available on the internet. It is much more organized, consolidated, well-written, and visually accommodating. It will remove the entertainment factor, noise, scrolling, clicking, advertisements, extraneous links, temptations, and other distractions you will find on an internet source, and will allow for quiet, contemplative, undistracted, and deep study. I came to know my calculus textbook very well. I know where everything is in it, and still refer to it 30+ years later. It carried me through my three semesters of calculus at UMass Boston. I found it to be well-written, organized, and graphically clear (*Calculus* by James Stewart, Wadsworth Inc., © 1987). I still keep all my homework assignments, syllabi, exams, and other items from those three calculus courses (as with my other college courses) in their respective folders.

Other Course Materials. Don't feel constrained to your course materials. This is what libraries are for. If you don't understand something after attending class or reading the book, make an honest effort to understand on your own. Asking for help is OK, but if you can solve the problem on your own via research and a dedicated, self-imposed effort, you will end up knowing the subject that much more. In addition, by making the personal effort to understand, and then asking for help afterwards, the help you receive will be a great deal more meaningful and clearer. I recall a very challenging engineering course at UMass Amherst that most students found to be very difficult. (I think we had one *A* in the course, seven *B*s, and the rest being *C*s, *D*s, and *F*s. I received a *B*!) The uncompromising professor was very clear about his expectations. Paraphrasing, he said, "*This course is not meant to be self-contained. If you don't understand something, before coming to me for help during my office hours, I want you to make an honest effort to conduct your own research using materials from the library and outside this course.*" Research will expose you to other authors with a different approach and method that will help you get a more thorough concept of the subject material. Information that is on the internet, including books, videos, articles, and so forth can be valuable, provided you're not looking at garbage. My claim is that in-print books—especially formal, academically-reputable textbooks—and periodicals that you will find at the library will be more valuable, professionally presented, and facilitate learning better.

Internet Sources. I don't want to discount all the math materials to be found on the internet, but as I'm sure you already know (or hope you know), there's a lot of junk on there. Just because something is on the internet doesn't mean it's correct. If your teachers and parents haven't conveyed this to you yet, they should have. Of the math materials that are on the internet, they tend to be:

- poorly organized
- sloppily presented
- hard to read and decipher
- incomplete
- distracting with irritating background music, ads, links, scroll bars, and sidebars
- not intended for the struggling student

I like the Kahn Academy due to its down-to-earth presentations and minimal distractions. (It has its deficiencies but I consider them to be minor.) One of the problems with legitimate internet sources is that they reinforce the bad temptation to be coddled and passively entertained. Moreover, students can't ask questions directly to the instructor. For more advanced math problems, try to stick to internet sites ending with ".edu" with reputable colleges and universities.

Textbook Reading Technique. Expect to be entertained by reading the textbook and expect to be disappointed. It's not a book to take with you to the beach on your summer vacation; you won't be reading a novel. Reading the textbook takes discipline to read and understand. Approach it with curiosity and a sincere desire to learn. Your self-discipline skills will improve over time, and it will set the standard for improved study habits for later in college, the workplace, and in your personal life. Don't limit your study habits to academics. Your good study habits will last a lifetime, regardless of your chosen profession or outside interests.

Here is the recommended technique for reading a textbook: Read the assigned/relevant section of the book three times:

1. Your *first reading* should be prior to the specific material being presented by your teacher. (This is one reason for obtaining the syllabus.) Do this the night before or prior to class during a break in a relatively quick, cursory manner without deep-learning intent. Your mind will be much more receptive to the new concepts once introduced in class.

2. Read after the class has adjourned along with your notes before starting your homework problems. Work through the example problems with the author. Take notes from this reading in your notebook. This *second reading* will make much more sense than the first reading. You'll certainly want to keep your book and notes handy to serve as a reference.

3. Read the book during and/or after completing your homework with the deliberate purpose of deep learning. This *third reading* won't take that long if you do it in a systematic fashion on a continual basis. The only time that the reading will be overwhelming is when you get behind and steer off track.

By the way, by accomplishing the three-time-reading steps above, you just partially studied for your next exam!

Pay Attention, Take Notes, and Ask Questions

Duty. Doing math is work. You may not be getting paid, but I can assure you a group of people (taxpayers) is paying to provide you the opportunity to be sitting there in school to learn. Make the most of this opportunity. Consider it your duty to pay attention to your math teacher, actively engage in the class, and do the required work.

Pay Attention. Teachers are human and they react positively to students who have good attitudes and a desire to learn. Students likewise benefit from teachers who pay attention to them. It's a synergistic relationship. Does your mind drift while in the classroom? Do you have an obsessive need to check your cell phone constantly? Fight it. Don't confuse the symptom of a problem (obsessively checking your cell phone) with the root cause. Despite what some in the school administration and pharmaceutical companies say, you don't need drugs, yoga, or meditation. You need to shut everything off including your cell phone, remove distractions in your life, and apply the F-word mentioned earlier (*Focus*). Your class probably lasts only 50 minutes—not a long time at all. Get sufficient sleep and exercise, eat a balanced diet of three meals per day, do what you are supposed to do, don't do what you are not supposed to do, and your ability to focus will improve significantly for that short amount of time.

If your teacher allows it, sit in the front of the classroom. Be the class "nerd" and disregard disparaging comments from the "cool" or popular kids. You'll figure out later in life that they were not cool if they disparage good students. My bold claim is that you can be cool, an *A*-student, as well as a good athlete, musician (or whatever your extracurricular interests are)—all at the same time.

Take Notes. You may understand everything your math teacher says in class but may only retain 10% of it, which is one reason that note taking is so important. You're making your brain work when you take notes. But taking notes unto itself is incomplete. You then must review and apply them. Ideally, this is done at your next break from classes as in a study session and/or right before working through the assigned math problems.

Note taking is a form of active learning. It's a skill; the technique improves with time. It's a necessary ingredient of homework and studying for exams. Date your notebook pages and write down everything that is important, not just what has been written on the board or displayed on a screen. Determining what is important and what isn't results from judgment that comes with time and experience, and a sincere desire for learning and self-improvement. It's part of the *skill*. Also record what your own thoughts were when something was said and its revelatory significance. Sometimes your teacher will also be helpful in guiding you with your note taking. I had a chemistry professor at UMass Boston who was very clear about his intent: "*When I write something on the board, consider it to be the word from on high. It will be on the exam.*"

A truism is that you can write something down that was said in class but still forget it. Contrastingly, you'll more than likely forget something if you don't write it down. To say to yourself, "*I'll remember that so I won't write it down*" may very likely turn out to be false. If you have problems relating to your teacher, note taking can serve as that surrogate.

Your notes do not have to be neat and clean; they just have to be understandable by you. My notes are probably incoherent to most people, but I understand them perfectly. I took my college notes quickly in my *fast handwriting* (translation: sloppy to the casual reader but readable to me) since I had no time to record them in a perfectly legible fashion lest I fall behind with what was being said in the lecture. I also included graphs, sketches, diagrams, and sometimes notes from classmates and professors who scribbled and drew in my notebooks too. This was done sideways, upside down, outside the margins…whatever it took. Upon review, I instantly recalled the conversation, relevance, and meaning from them, and not only from a class just completed, but years later!

Ask Questions. Asking questions is also an active learning component. Not only does it answer the legitimate questions you—and others—may have, and keep the teacher on her toes, but it also helps you keep alert and engaged. When questions arise from your fellow students, dismissing them contemptuously and with exhaustive disdain is easy, until you get surprised by the answers. Mutual respect works wonders here. When asking questions to your teacher, do so tactfully, respectfully, and with appropriate timing. Your deference, politeness, sincere interest, and good attitude will be appreciated, and will open up others' minds, especially when you or they hear the unexpected answer. Asking questions also allows you to hear yourself, which can sometimes be embarrassing, disturbing, but also enlightening. This is all part of the learning process. Asking questions, and listening to those of others, also presents a good opportunity for note taking.

Your teacher may certainly react with a degree of dismissiveness from time to time. (Remember that teachers are human.) React not with your own brand of disregard, vindictiveness, and an attitude of "*Never mind*" or "*Forget about it*", but by following through tactfully, explaining the reason for asking the question, and by elaborating. This will improve with time. Empathy with the teacher helps.

Accept some mortification when you ask your questions. Swallow your pride. Make a fool of yourself periodically. You'll be surprised at how many other students don't know the answers, alternate approaches, or have the same level of frustration or confusion as you do—including the so-called "smart", overconfident ones. Think of this as a learning opportunity. And take notes of the feelings you experienced. Again, upon note review, you'll be instantly brought back to the moment when you asked the questions and will know the subject matter that much better. You'll be engaged and active in the class—and the math!

Amalgamation. I've always done my best academic work in classes delivered by organized, disciplined teachers and professors. Correspondingly, I've done my worst with those who tended to be the opposite. One of my high school math teachers fit the latter description. She was very kind, likable, and competent, but lazy and not empathic in that she always used an overhead projector with those plastic slides to write

on instead of the chalk board. (These things may predate you.)
I stopped going to class after a while due to my frustration
and irritation (among other emotional things going on) and
with the false assumption that I wouldn't need this kind of
math in the Marines—my concrete goal. Of course, I ended
up with *D*s and *F*s, and a sharp admonishment from her one
day when she cornered me in the hallway between classes.
She was right, and, in hindsight, I should have followed my
own *current* advice (I didn't know at the time!) to have made
up for her deficiencies and my own shortcomings. That is,
I should have used other means covered in this chapter to
have picked up the slack. Switching classes and teachers was
too late in the academic year by the time of my not showing
up. My disillusionment, ignorance, and lack of confidence at
that time with math prevented me from paying attention,
taking good notes, asking questions, and doing the assigned
homework. Not surprisingly, I didn't have much self-
assurance in asking questions, although I never would have
admitted to it then due to false pride. The other students
in the class had no problems that I was aware of. *I was the
problem.*

In a somewhat analogous situation at UMass Amherst,
I registered for a required mechanical engineering course
where the professor was a disorganized, undisciplined
blowhard—at least to me—who was teaching the course
as he saw fit. By this time in my life, I was fairly wise to
the world, and I could see through his nonsense. I switched
classes to a more competent professor immediately. It was
a prudent decision since the class material learned with the

more capable professor has been very valuable to me in my work life over the years.

My advice is to do the same. If you're not connecting with your teacher, recognize it early and switch classes, if possible. If you can't switch, accept the situation, stick with it, and use alternate methods to make up for the teacher's shortcomings.

Do the Homework

Where the Real Learning Happens. If you're a typical student, you don't look forward to doing homework, specifically math homework. There are other things you'd rather be doing. However, as you know, homework is a necessary component to academic success. Homework is where the real learning occurs, which is why it is assigned. But homework must be approached and executed correctly to reap its benefits.

> *Learning is not attained by chance, it must be sought for with ardour [sic] and attended to with diligence.*
> - Abigail Adams, wife of President John Adams and mother of President John Quincy Adams, 1744-1818

In a classroom environment, the learning process is structured and guided. Concepts are elaborated on by a human teacher. You share your frustrations, pain, suffering, and joy with your classmates, some, of whom, are your friends. Homework is different. You're on your own. Self-motivation is the driver now. You've been given the necessary materials.

Application and learning are up to you. But what is homework in the context of math? It's more than just doing math problems, which, of course, is critical. But doing your homework is also a nonobvious way to study for exams. You're also reviewing, recalibrating, developing logic and reasoning skills, learning how to conduct research, and keeping your mind sharp. When you can discuss your homework problems with your classmates, you're hearing how to approach problems from a different perspective and realizing your own mistakes, absurdities, misinterpretations—and triumphs too!

Purpose of Homework. Homework is obviously done outside of the class. This is where deep learning and understanding happens. The classroom can be entertaining and dangerously *passive*. You can appear to understand something in the classroom but later realize that you didn't truly understand the material when the time comes to complete the homework. Reading the book, reviewing your notes, and doing math problems is *active* learning. Here, you are not being supervised or coached. You can't just pay attention and nod your head. I was still foolish enough in this regard at age 27 when taking a course in statics (another math-intensive engineering course; not be confused with statistics). I paid attention in class, read the book in a cursory manner, understood the concepts clearly, but only made a half-hearted effort at doing the problems. The professor didn't collect and grade our homework assignments, which allowed me to be too cavalier in my approach. We went over them in class together where the students graded themselves.

Sadly, I flunked the first exam but no one else did. The problem was me again. Luckily, the professor disregarded each student's lowest exam score (except for the final exam) so this one didn't count. I was humiliated and angry at myself. My shortcomings were made obvious and I never made that mistake again. Brains need to be exercised constantly with hard problems on a regular basis during academic semesters to learn properly.

Recognize this now: learning is rarely a straight path. The formal education you are receiving now in classroom-type environments guided by teachers is *straighter* than On the-Job Training (OJT; in other words, real experience) that you'll receive in the work world. Certainly not everything can be learned in the classroom, and not everything can be learned with OJT. We need both types of education. When you are studying and doing your homework, and you are challenged, frustrated, and want to give up, you are learning. What you do with this opportunity will help determine your fate. We learn many times by failure, not by success. This is a part of life. Understand and accept this now.

> *Experience keeps a dear school but fools will learn in no other.*
> - Poor Richard

Method. Math is not a spectator sport, as many have said. Your brain needs to be engaged to make the learning process work so that you can succeed in math. The burden of your learning is mostly on you—not your teacher. As I mentioned, reading the book and reviewing your notes are part of doing your homework. Do not separate these from doing math problems; they need to be grouped together. Doing math problems is a form of inculcation, in other words, learning by frequent repetition. Inculcation applies to writing, sports, music, public speaking, and just about all other learning areas. It is a critical and necessary part of learning.

> *Many times, the hardest part about doing homework is starting it.*

"Professionalize" your written homework assignments to empathize with your future self and minimize exam-studying frustration, since your professionalized homework assignments need to become part of your exam-preparation material. Here's what I mean by professionalizing your homework, regardless of whether or not you're expected to pass in your assignments:

- Use a mechanical pencil coupled with a separate eraser
- Put your name on your homework assignments and date them
- If you must use preestablished worksheets, but run out of space, attach your scratch paper that is also appropriately labeled
- Put the corresponding problem numbers on your scratch paper in an organized manner

- Show all your work and be methodical about it so that your teacher and your future self can follow your reasoning
- Write clearly
- Organize, align, place in numerical or chronological order, and simplify all your work, so that even your scratch work and sketches are situated in the right areas
- If you're not required to pass in your homework assignments, but they're to be graded and reviewed together as a class, such that the onus is on the students to correct them, use a red pen to make the corrections and include notes and admonishments to yourself where you erred
- Save your homework assignments and place them in your math class folder in an organized fashion (recall class folders from the *Location and Class Materials* section)
- Constantly reevaluate and improve your homework's organizational method

What if your teacher doesn't assign you any homework, or the homework is too easy (that is, your teacher is pandering to you)? If you learn of this practice in your first few days of class, drop the class and transfer to another more difficult class immediately (if possible, of course). Your teacher may adhere to this policy for more than one reason. Maybe he has no confidence in you students to do the work, so collecting and grading homework is not worth his time. Or perhaps your teacher is making homework unchallenging in a desire to be liked. Nevertheless, you're missing out on an education and an opportunity that may never surface again. Not surprisingly, I also hear a lot of, "*I don't have any homework*" from students from my tutoring sessions. This may be true for *assigned* homework. The obvious remedy is to assign homework to yourself. (Recall references to the syllabus and scheduling covered earlier.) Schedule your self-assigned homework for 30 minutes per day minimum—even on weekends. Once it's done, you'll be much better off, and will have forgotten about the "pain and suffering." If done in the morning, the rest of the day will be yours.

Prepare for Exams Properly

How *NOT* to Prepare. Let's begin by covering how *not* to get ready for exams, especially by cramming. Cramming the night before and at the last minute is no way to study and is *usually* unnecessary, provided you've been actively engaged all along, including the reading of the book and doing the homework problems (in other words, doing what you were supposed to be doing during the course). If you cram extensively and routinely, then you haven't done your job. By the way, I'm not discounting cramming. It has saved me many times. But what I am saying is to avoid it and the undesirable side effects such as self-loathing, despair, anger, lashing out at others, and lacking a sense of control. Reserve cramming only for those last-minute, unavoidable, legitimate, unpleasant events of life (for example, work, conflicting deadlines, family emergencies). Many can get away with cramming routinely and pass the tests, but there's no deep, long-term learning. This is called "winging it" and living by your wits. These people eventually become know-nothings with college degrees in the work world. (Unfortunately, they're quite common in industry, and some even hold high offices in government.) Don't be impressed by them or follow their example. They will break eventually in one way or another, even though this could take a while (a long while in some cases).

Reap the Benefits of Homework. If you did your homework and followed all the other tenets of this chapter, then the hard part about studying for exams is essentially over. The rest should be relatively easy. You already did the hard work. You read the book, did the homework, took good notes, reviewed them regularly, absorbed the material, and your thoughts and subject matter have been organized in your brain. But you still need to review everything and prepare for the exam. Start a few days in advance, if not earlier. *Notice that I didn't say the night before!!*

Condense Your Notes. Let's say you accumulated sixteen pages of notes in your notebook that cover the exam material based on a course in algebra and trigonometry. These sixteen pages can and should be read, consolidated, simplified, and condensed ultimately into one page. How is this done? First, read over and understand what you wrote in your notebook. Second, get a blank, unlined piece of paper, draw a line down the middle such that you're left with two columns. Third, begin transposing the key, important notes from your notebook into the paper starting at the top left using small but readable handwriting (readable to you). This includes key diagrams, terms, self-criticisms, and so on. Use shortened words and abbreviations that make sense to you. Omit the clutter, repeats, incoherent scribbles, unimportant things, etc. Once you've covered a particular topic, draw a horizontal line in that column below what you just wrote and repeat for the next topic. Continue this process until you're done with the notebook content. See Figures 2 and 3 for schematic illustrations of the process. If you exceed the one page, put your notebook to the side, repeat the process but use the newly created condensed notes sheets to transpose from until you can crunch everything into one page on a new sheet. This writing, rewriting, reorganization, and consolidation is a lot of work, but it forces you to think, read, comprehend, and remember. By the time you're done, you know the material pretty darn well. Now read over your one page of condensed notes and make sure you understand everything. This technique improves with time.

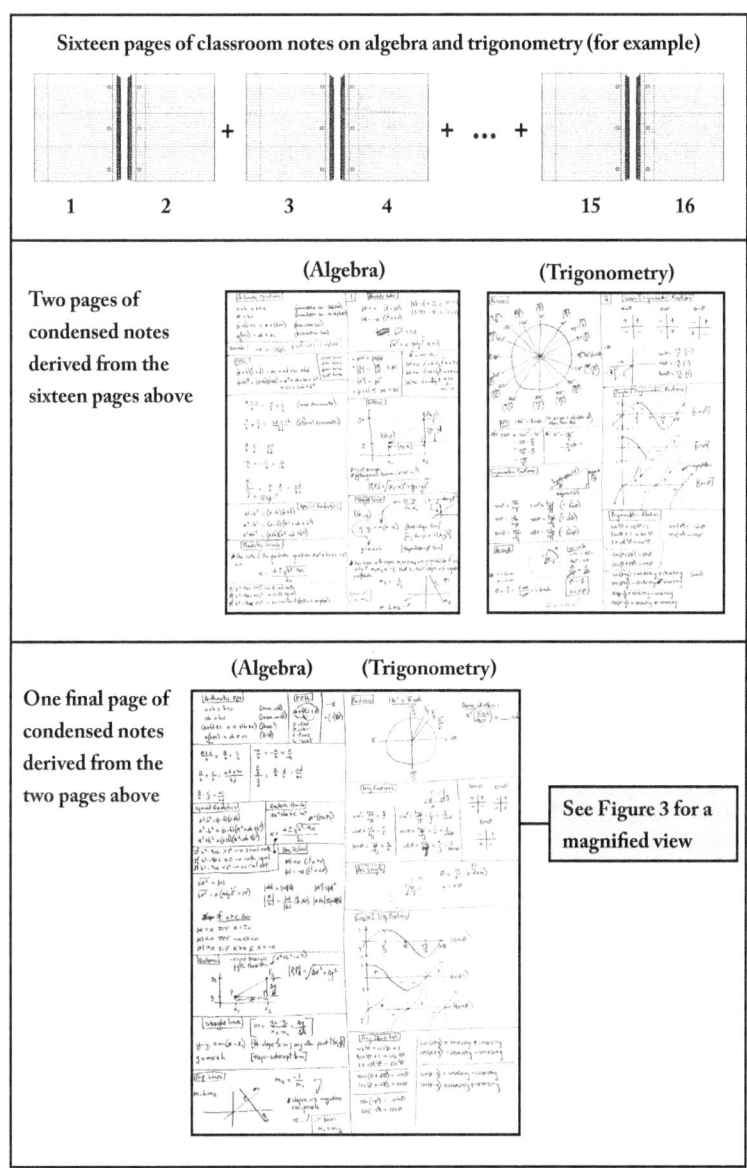

Figure 2 – Schematic of Condensed Notes Process

Summary:
- Sixteen pages of notes condensed into one
- Readable and understandable to you
- Results in deep learning

(Algebra)   **(Trigonometry)**

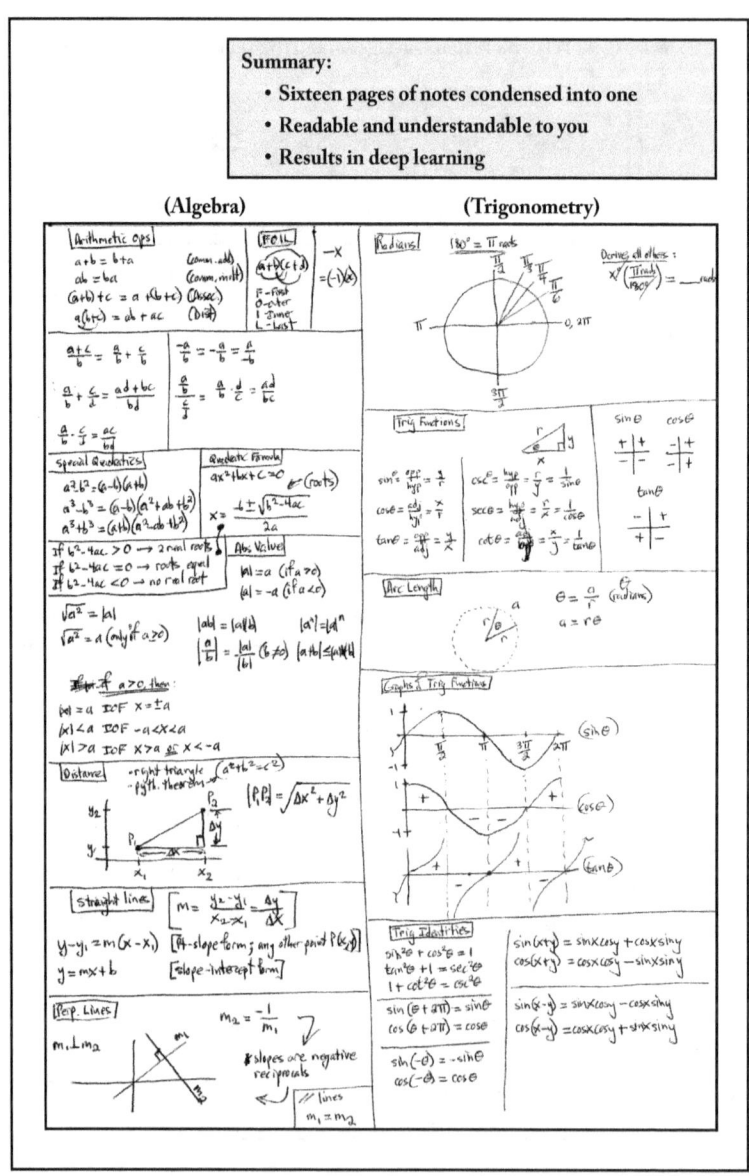

Figure 3 – One-page Result of Condensed Notes Process

Review Previous Exams. We already know that in math, one concept is built on another. We cannot take and pass a test then forget about the material covered. Therefore, reviewing previous exams is necessary (assuming you had one or more in your current math course). Review the material certainly, but also recall your mistakes, misinterpretations of questions, teacher's comments, and emotional responses.

Review Your Homework Assignments. Since you professionalized your homework assignments and kept them organized, now is the time to review them all. Like your previous exams, you'll want to review what you did right, where you went astray, what you were thinking at the time, teacher's notes, the subject progression, and, of course, the math.

Do Refresher Problems. Doing new refresher problems (those that you haven't done before) with known answers for checking is extremely important and a very active part of studying for a math exam. This is the pinnacle of your studying efforts. Other than the exam itself, nothing is more revealing about how well you know the subject than when you have a new problem and a blank piece of paper. Do as many problems as time will permit but do them strategically. Your time is finite, so do problems that cover each topic that will be on the exam.

Exam Preparation Summary. Constant, persistent, effective, and efficient work habits during your math course leads to minimal exam studying, reduced anxiety, deep learning, and high exam scores. The bottom line is that if you stayed engaged and did what you were supposed have done during the entire course, you've mostly prepared for the exams prior to the formal exam studying process. And in a field such as mathematics, where you will not have the luxury of passing an exam and forgetting about the subject just learned, this deep-learning habit will work wonders for long-term retention and execution.

Change Your Attitude and Approach

Changing. Adopting the correct attitude, approach, and way of thinking is arguably the key part of this chapter and book. The other principles in this book are meaningless otherwise. Success or failure will depend on your attitude and approach and how they relate to math, as well as just about everything else you will encounter in life.

Attitude. *Attitude is everything.* I hope that this is not the first time you heard this phrase. If it is the first time, then burn it into your brain. The right attitude is necessary in school, the work world, interpersonal relationships, sports, other extracurricular activities, the military, and, well, everywhere. But what is the right or a good attitude? I could write another book on this subject but here are some of the key traits (there are others, but you'll get the idea):

- Enthusiasm
- Strong work ethic
- Humility
- Helpfulness
- Persistence
- Self-reliance
- Patience
- Responsiveness
- Communicativeness
- Dependability
- Resoluteness
- Resilience

- Self-discipline
- Getting along with others
- Awareness of self-weaknesses with a willingness to improve
- Empathy with others
- Taking criticism without being vindictive
- Acceptance of difficult assignments and viewing them as learning opportunities

Approach. To steer your approach, let's review what you want. Sparing details, you want to live to your potential and, as I alluded to earlier in the book, you want options. Therefore, you must study with a purpose and to *learn for life*, that is to say, not for the short term as in getting through your current math course but for the long term. This piece of wisdom surprisingly came from my high school math teacher I described previously whom I had trouble with. (She constantly used the overhead projector.) As you may have predicted, I ignored her guidance at that time, but I didn't forget it.

You will learn only if you *want* to learn. I discovered this in the Marine Corps. Some Marines came into the Marine Corps as derelicts and left as derelicts; they had a bad attitude and never adjusted. Most others did learn something from the Marine Corps' virtues and adopted them for life. How you emerge from high school will be up to you.

Another philosophy to consider is that if you are self-satisfied, confident, relaxed, in control, and comfortable, then you are not living life correctly. We all need to accept problems, aggravations, and setbacks, and be thankful for them. *What doesn't kill us makes us stronger.* Seek out difficult tasks even though you may fail at them. Consider them challenges and opportunities to learn and grow. Every time that I get significantly disappointed in life when things don't turn out the way I had planned, I find out later (sometimes much later) that my disappointments were blessings in disguise.

Let's use a sports analogy. We hear this all the time: "*I'm not an athlete.*" Also, "*I'm terrible in basketball but I'm good at tennis.*" You don't have to be in the National Basketball Association (NBA) to play basketball, or in the Professional Golfers' Association to play golf. If you're skilled at tennis, then you must have some athletic ability such that you may not play basketball so well because you don't play it regularly, you don't practice it enough, the coach never lets you play to allow you to improve your skills on the court, or you practice wrong. Here's a myth (for 99.99% of us): Work hard enough and you will be as good as NBA star LeBron James. Here's a fact: Work harder and correctly and you will get better, but we can't all be as good as LeBron James. There will always be someone better than you in something. Learn to accept this. Just keep in mind that you have talents that LeBron James doesn't have. Win graciously and lose graciously. Get involved with sports for the right reasons, and not just for physical exercise. Sports offers so many other intangible benefits such as social skills development, athletic improvement, healthy competition (*Competition betters the breed*), enjoyment, recreation, living to your potential, and mental health. There are plenty of others. I play guitar, basketball, golf, and also bike and run. I will never be great, or what I consider to be acceptably proficient at any of them (partly due to time constraints, partly due to lack of natural talent, and aging does not help), but I do them anyway because I enjoy them, get mental and physical benefits from them, see the payoff with practice and perseverance, and realize the *indirect* investment returns to work and life, in general.

Now translate this approach to math. Do it for the right reasons and open your eyes to the obscure. Sometimes an unbiased, outside critic can help you illuminate the nonobvious. With math, your critic can be your teacher and other qualified persons—and your grades! They may tell you things that are unpleasant. Accept their honest, legitimate criticisms (pray that your critics will be tactful) and learn from them.

Work Ethic. Ignore and disdain the easy and glamorous life and avoid those people who advocate for it, and adopt a disciplined, purposeful work ethic. Of course, this is easier said than done. We all know intuitively what a good work ethic means, but putting the definition in writing is not so simple. Here are some generally accepted characteristics that apply to both physical and mental labor (yes, there's an overlap with traits of *attitude;* they're interrelated):

- Diligent
- Accountable
- Responsible
- Dependable
- Self-disciplined
- Self-controlled
- Self-motivated
- Initiative-taking
- Persistent
- Resilient
- Purposeful
- Energetic

These adjectives are the goals we all wish for. They're so much easier to read, write, and say than to emulate, but they are virtues to strive for.

Adopt the attitude that you've been given a tremendous educational opportunity and that you are in school to learn, not to socialize, have fun, and be entertained. You're at school for six to seven hours per weekday, which isn't that long. Work the entire time, study during formal study times and other downtimes, yet allow, of course, for lunch and healthy time-checked socializing. The opposite is unhealthy mischief and extremely wasteful time spent on social media among other things. Recall from the *Get Organized and Manage Your Time* section to not goof-off and squander this educational chance that many others in this world do not get.

I'll never forget my time at NAS Memphis during *Avionics Technician Course*. This was an unusual course where we Navy and Marine Corps students (maybe 100 in a room) had to sit in individual carrels[4] and quietly read specific modules in books, understand the material, perform the calculations, conduct a lab exercise, then take and pass a computer-managed test before progressing to the next module. Our job was to conduct this self-driven style of studying six hours per day with 10-minute breaks per hour. (Other duties preceded and succeeded school time.) Our "instructors"—senior noncommissioned officers—were there to answer any questions we may have had from the course materials, as well as to maintain discipline. If a student was not progressing,

---

4   Adjoining desks partitioned with small barriers

the instructors were made aware by the computer-generated reports. We were also given negative incentives to progress. For the Sailors, the instructors semi-jokingly warned them that they would be transferred to the *USS Neverdock* (Get it?) as boatswain mates if they failed to perform. For Marines, they let us know that there were many job openings in the infantry. *Avionics Technician Course* is where I learned that long, exhaustive, quiet, difficult study does not kill a person. Nobody died during the entire time I was there, and nobody's brain hemorrhaged or blew up, although we did have some nonperformers who were shipped out to who knows where.

> *Nothing in the world can take the place of Persistence. Talent will not; nothing is more common than unsuccessful men with talent. Genius will not; unrewarded genius is almost a proverb. Education will not; the world is full of educated derelicts. Persistence and determination alone are omnipotent. The slogan 'Press On' has solved and always will solve the problems of the human race.*
> - Calvin Coolidge, 30th president of the US, 1872-1933

I provided the steps in studying and learning math. Your job now is to do them—all of them, not some of them. Put in the required effort. Make it a regular habit. Stay engaged and make your brain work. Don't stop daydreaming (it can be healthy and useful) but keep daydreaming in its place. Work hard habitually so that your work becomes easier in the end and in the math courses that follow. Math talent certainly plays a role here, but persistent laborious work pays off more. Let me use another sports analogy to get my point across. A person is a good athlete. But that person would be a better athlete if he/she: practiced more; practiced correctly, efficiently, and effectively using the proper techniques related to the relevant sport; ate better with proper nutrition; rested strategically and obtained adequate sleep; balanced other facets of life; scheduled his/her time; and so forth. World-class gymnasts, figure skaters, public speakers, and musicians make their performances look so easy when in fact, their performances are based on years of painstaking work. Don't overlook challenging yourself either. Think of a wrestler who during practice sessions intentionally wrestles with bigger, heavier opponents. When the time comes to wrestle in his weight class, the match is approached with confidence and fought with relative ease. The same holds true with musicians and in other fields. If you want to play the trombone better, play with professionals. You may not be able to keep up with them, but their techniques and expertise will wear off on you—consciously and subconsciously.

I mentioned physical and mental labor earlier. They both matter and are important. Both require some degree of toil, problems to overcome, frustrations, successes, and failures. Speaking of which, success is a great incentive, but as I said before, we learn by failure many times. Persistence pays off. But judgment is needed here, however, which comes with time, maturity, the making of many mistakes combined with the right attitude and approach in learning from mistakes.

This attitude, approach, and judgment also applies to math, your other courses, and life, by and large.

I view many (not all) school/academic-oriented TV shows that are geared to kids to be patronizing, intelligence-insulting, immature, colorful, all peppered with light humor. The message seems to be that academic learning is fun, straightforward, entertaining, and easy. (How else would they get their ratings?) From my perspective, these shows send the wrong message. Learning, with an emphasis on math here, is not always fun and interesting, and it's not going to be many times. Sometimes it is, but much of the time it isn't. It can be downright difficult, complicated, and fraught with drudgery and frustration. You can make the work interesting with purpose, as well as with your attitude and curiosity. These TV shows omit the pleasure derived from the payoff in persevering work and solving complex, baffling problems. I also find these shows to be hosted by some nerdy guy talking foolishly and childlike with happy, clean, well-adjusted, well-dressed, well-fed, non-frustrated kids. This is not new; they were like that when I was school-age. My take is that for those students struggling with math and other subjects, these shows are not helpful and exacerbate the problem.

Discouraged Math Students. What are my words of wisdom if you are discouraged with math? There's no *easy* answer, but there are answers—they're just not easy and may not be so agreeable. Recover, adapt, pick yourself up by your bootstraps, and follow the tenets of this chapter. Let me go on.

Positive activity in times of despair breeds hopefulness. This means returning to math fundamentals. In my tutoring experience, troubled math students have problems not with the so-called *advanced* math, but with basic math. I recall one troubled tenth-grade algebra student I was helping some years back. She was on the verge of either receiving a poor grade or flunking. She just didn't get it and was very distressed. After working with her for about five minutes, I discovered that her problem was with some very basic math. I explained her misunderstanding and we did many problems. I explained that the teacher was not wrong (by examining her notes) but was skipping steps that she was not picking up on. The teacher was doing math in his head without properly explaining himself to the class. Once we identified and fixed the problem, she went away, came back a week later, then showed me her grade from her latest exam: 99%. (I asked her why she didn't get 100%. She made a simple error.) A similar scenario happened with a twelfth-grade calculus student I was working with. Again, she wasn't doing well in her class, but her problem became immediately evident to me in that she was having trouble dealing with the manipulation of positive and negative integers, as well as solving very basic algebraic problems; that is, some math that she didn't learn thoroughly in earlier grades. She had no problem with the advanced concepts of calculus; it was the basic stuff that was killing her. These problems were not insurmountable. They were not too late to fix. However, some extra effort on her part was required to catch up.

Don't succumb to the self-fulfilling prophecy of, "*I can't figure this out, therefore, I must be dumb*" or "*I can't do it so I'm going to stop trying.*" Falling into this trap is effortless. I've done it multiple times. On the other hand, don't be self-satisfied, overconfident, or complacent when you succeed. And please don't live by your wits, pass the test, and declare victory. This is a prescription for disaster.

Self-pity is normal, understandable, and easy, but it won't solve the problem. Just like spilling milk at the dinner table, the milk runs down the side of the table and onto the floor, and you are sitting there angry, frustrated, and embarrassed. You can't walk away, and you can't continue eating and socializing. The mess must be cleaned up immediately and thoroughly or stains, spoilage, and odors result. You're not in a restaurant where the staff will clean it up. You must solve the problem and clean it up yourself, and clean it up now. There is no choice. Translate this into your math problems. They can be solved, and you can solve them by studying correctly. And they must be solved now. Putting math problems off "until you have time" won't work. There is a bright side to all this. You will start to see the positive results immediately (perhaps on a small scale at first, but they will come).

Be wary of tutors and teaching companies. I'm not discounting them, but if you choose to go down this route, ensure your tutor or teacher is *qualified*. What I mean by this is that your tutor or teacher has the ability to empathize with you. Had I sought this kind of help while I was in high school, it would have been useless to me given my independent streak, steely goal of joining the Marines, and unfulfilled curiosity to know what math is, what to do with it, and why it's important. I don't believe anyone would have given me satisfactory answers at that time. (No tutor or teacher could have empathized with me then.) Many times, assuming tutors and teaching companies are not imposed on students by their parents, you may *want* their help but don't *need* their help. Don't get *want* and *need* confused; they're not the same thing. Sometimes all you need is a book like this (I'm partial, of course) or a mini explanation on a key point. Just keep in mind that to learn math properly you still have to do *work*. To illustrate this, consider free online academic courses, not the short videos you often see on social media, but full courses offered by colleges for free on a variety of subjects. Motivated students fare well with these, but unmotivated students who expect to be entertained, learn by "osmosis", and not do any work typically perform poorly. These courses are not designed for *passive* entertainment. The lecturer is presenting material, but the student must then follow the lecturer's guidelines and complete the assignments (homework must be done) to get any true lasting value out of the courses. This is *active* learning.

Classroom Dynamics. Classrooms can pose an intimidation problem. When you don't understand something that the teacher is covering but everyone else appears to, a reluctance to ask questions is natural (even well into adulthood). You can easily say to yourself, "*They get it, and I don't. There must be something wrong with me.*" Asking questions can be awkward, intimidating, embarrassing, and demoralizing, especially when you do it repeatedly.

Here are some words about those students who get it when you don't. First, this book is not for them. For them, math can be entertaining and there's nothing wrong with that. Let them knock themselves out. They might find it entertaining because they study math properly and are prepared and, hence, successful at math. These students may also come from academic-supportive families, which you may not have. Of course, there are some students who are highly intelligent and/or can think in the abstract better than most. They have a natural talent. Don't forget though that some of these students don't know what math is, what to do with it, and why it's important; they just accept it and thrive, at least in the math context. I have seen many students thrive in high school only to peter out later in college. Expanding, the same holds true for those who excel in college and later flop in the workforce. School is not everything. Accept the fact that there will be those students who will be more talented than you are at math, but that your math trouble (the *Problem*) is probably your study habits (the *Cause*), so follow this chapter's prescription on how to study and learn math correctly (the *Solution*).

Life Is Unfair. Let's face it: life is not fair and is not meant to be easy. We have to remind ourselves of this fact regularly lest we start drowning in self-pity, never mind succumbing to envy of others. Chances are that you can see, walk, run, and think for yourself. Not everyone can. Moreover, you probably have freedom, opportunities, reliable food sources, and access to an adequate supply of potable water where much of the world's population does not. We all have abilities and strengths that others don't—including those with a gift for mathematics. You may not discover the talents that you've been blessed with until later in life, but you have them.

Don't squander this opportunity—that many kids in this world don't get—that we call *school*. Your school time is limited and expensive; it's an opportunity that will end upon high school graduation, where the dynamics of future schooling will be much different. I recall a radio interview (I don't remember the subject) from a few years ago where a father was describing his semi-joking admonishment to his daughter for her comment about something in which she said to him, "*That's not fair!*" His response was, "*You should get on your knees and be thankful that life is not fair.*" In other words, his true meaning was that his daughter was not living in a war zone, a totalitarian state and not subject to rape, killing, maiming, destruction, kidnapping, starvation, torture, and unjust incarceration. That she should be so blessed. Is that fair?

Parents or Guardians. Your home life is probably OK since you are doing alright in your other courses, but what if it isn't? What if your *real* problem with math is due to something else? It could be a number of things. One problem may be your parents or guardians. These people could be your biggest academic enemy, where you could be paying a price with poor math grades. Certainly, your home life affects you emotionally and, consequently, your academic performance.

Ideally, your parents or guardians are providing you with food, shelter, clothing, and moral guidance. They are just, calm, stable, virtuous, and academically supportive, and started you off on the right foot in your formative years. Their support is constant from your birth until you are 18 and beyond. Yes, this situation would be ideal, wouldn't it?

Well, family life is never ideal and not the same as it was when I was growing up in the 1960s and 1970s. But let's be clear: it wasn't ideal for many then either. However, most (not all) households had two parents, where there was one main breadwinner, who was typically the father, and the mother, who did everything else, if not more. She was the main child-rearer, housekeeper, errand runner, disciplinarian, nurse, and you name it—at least during her children's initial years. She may have had a part-time job. Her career options were much more limited in those days. She was the one who kicked us out of the house when we were driving her crazy and said, *"Go outside and play with your friends, and don't come back until dinner time!"* Some parents were monitoring their kids' schoolwork, others not so much.

Today we see a lot of two full-time-income homes, single-parent/guardian households, and hectic go-go, constantly connected, overly entertained lifestyles. In some homes, the routine is all work and no play, while the opposite is true in others with the allowance of the wrong types of freedoms. If kids get time to play, it can be structured and scheduled as with play dates and sports. When parents or guardians are always working, this means that they are not at home when the kids get home, and maybe too tired to pay attention to anyone once they arrive.

Academically, parents or guardians are many times deficient in math (never mind science) making their math support ineffective. Parents and guardians are notorious for telling their kids, *"Do as I say, not as I do."* In other words, *"Do your homework while I sit here for hours on end watching TV."* They can also have bad attitudes and unreasonable expectations of teachers reflected in statements such as, *"It's the school's or the teacher's fault that my child isn't learning",* or telling the teacher, *"I work in the real world and you don't. Therefore, I know better than you."*

You've probably already seen parents or guardians with bad attitudes that ruin youth sports. They can also be overly engaged; these are the so-called "helicopter parents" hovering over every aspect of their children's lives. On the opposite end are the disengaged or excessively permissive parents or guardians who offer no academic or emotional support whatsoever. They may be the type who are constantly running around for the sake of running around, hiring others to perform routine home maintenance tasks, and eating out at restaurants consuming unhealthy food much too often, in lieu of home-cooked, nutritious, sit-down meals at the dinner table with no electronic devices being on.

Instilling a proper work ethic—both physical and mental—in you is tough for them to do if they don't have it themselves. Doing homework in a chaotic, noisy, and undisciplined household is not easy. (For some of you, the situation may even be worse. I'm well aware of what can go wrong in a child's life with physical abuse, mental abuse, or extreme neglect. You have my sympathy if this is your situation.) If your parents or guardians are not helping you develop your work ethic by fairly assigning chores, ensuring you complete them whether you want to or not, teaching you how to do things properly, implementing reasonable limitations on your freedom to keep you out of trouble, and, most importantly, setting the example, then they're doing you a grave disservice.

What You Can Do About It. As they say, you can choose your friends, but you can't choose your family. If you live in a less-than-ideal household, don't allow your situation to be a "legitimate opportunity" for self-pity and surrender. Like the spilled milk, your math class obligations and resulting grades can't be wished away. Remember that your time in school and childhood is temporary, but so is your opportunity for a "free" (taxpayer-funded) education. Since life is unfair, recognize and accept the unfairness and nonideal conditions and deal with them as best you can. Determine what you can and cannot control. There may not be many choices. The upside is that you'll be 18 soon (if not already) and you'll then have more freedom.

You have influence with your family members; don't think that you don't. You can use subtle, unoffensive, tactful suggestions to refuse or raise objections to activities that you know are counterproductive to your academic success and well-being. Make reasonable demands, and also try to empathize and understand everyone's problems and weaknesses. Everyone has them and your parents, guardians, and siblings are not exceptions. What is the best way to get someone to do something? Set the example. This works best from the top down, but it can also work from the bottom up where subordinates surreptitiously influence their superiors (that is to say, children persuading their parents or guardians). I can't guarantee that this method will work. In case it doesn't, you can also seek further guidance from trusted teachers, coaches, school officials, and other adult mentors.

Friends. Contrastingly, you can't choose your family but you can choose your friends. Ideally, and analogous to parents or guardians, you want fun, physically active friends who are enthusiastic about school and learning, who do their homework and stay out of trouble, and have a positive effect on you. You need friends but be on your guard. Your friends can influence you (and vice versa)—consciously and subconsciously. The wrong friends have the capability of wearing you down and being negatively persuasive. In a worst-case scenario they can get you in trouble with the law. Like it or not, people judge a person by the friends he or she keeps. Hanging out with the wrong friends can cost you financially, employment-wise, and otherwise. To use a business metaphor, one of the main principles of marketing applies here: *Perception is reality.* Be cognizant of the temptation to join the crowd, especially the type that extols or advocates laziness. *Don't hang with the gang,* and *If you hang with trouble you get into trouble,* are maxims to live by. Problems with bad friends or friends with poor judgment are exacerbated by the advent of social media. Keep in mind that what's posted online may stay there forever. (Thank goodness social media didn't exist when I was young!) A friend of mine once admitted to me that he and his other friends (they weren't bad kids by the way) would make fun of one of their mutual friends (playfully, in this case) about him studying hard and doing well in school, and later majoring in a STEM[5] field in college. My friend is not doing so well today financially, whereas his friend who applied himself in school is thriving—financially and in other ways. On the other hand, chum around with the industrious, enthusiastic, and hardworking types, and their traits will wear off on you. Enthusiasm is contagious. Become friends with those who will do you good. The best way to meet new friends is through extracurricular activities such as sports, school clubs, the band, and so forth. I'm still friends today with those I knew in elementary school, junior high school, high school, and the Marines, and yes, most are good friends.

General Self-improvement. There's something wrong with all of us. We need to be self-critical and honest with ourselves enough to identify our follies and to work on self-improvement for our entire lives. This requires constant self-evaluation, and the judgment that comes with experience. Accept the fact that your study habits, attitude, and approach are evolving and need improvement by default. They key is to not work harder, per se, but to do so more efficiently and effectively. The person, be she a musician, athlete, or math student who wants to get better, who works hard at getting better, and is open to constructive criticism usually gets better. None of us will ever achieve perfection, but if you work at self-improvement correctly and constantly, you will

---

5   Science, Technology, Engineering, and Math

improve, gain experience, learn lessons, and be the happier for it. All of us are weak in some ways. Accept this but do something about it. We tend to be emotional creatures, not logical ones, but we are capable of making decisions and being resolute. The process is not easy. You may never shake your weaknesses, but you can certainly try to recognize them and minimize their negative effects.

Write out what you want—and don't want—out of school and life, in general. This sounds easy, but it's a difficult task for most people. (It can get harder as you get older.) The process could take five minutes, five weeks, five months, or five years in some cases. Putting your goals on paper will increase your confidence, make you more resolute, provide insight into your own thought processes, and help increase the execution part (the tough part). Make this a periodic habit, like semi-annually or annually, or during school breaks and refer to your previous writings (save them). Adjust as needed. Many successful people go through this exercise. From Socrates[6], *"The unexamined life is not worth living."*

In addition to giving yourself honest criticism, also accept genuine, constructive criticism from qualified outsiders. Their criticisms may not be easy to take. Expect some anger, resentment, brooding, and embarrassment initially, and avoid vindictiveness at all costs. After these feelings subside, determine if the outside criticism is correct and warranted, and then go to work.

---

6   Ancient Greek philosopher, circa 470-399 BCE

Putting the painful criticism aside, seek out the wisdom of others—particularly academic wisdom. Just be forewarned that some "wisdom" could be erroneous or only partially right. (I don't think mine is, but I'm biased as you know.) To alleviate this problem, seek wisdom from many, not a few, and learn from your current knowledge and experience. Seeking wisdom from others doesn't necessarily mean asking for it directly, but it could mean asking a lot of indirect questions, as well as simply reading, listening, and watching. Another means to gain wisdom and judgment is to learn from others' mistakes. You'll learn plenty. Furthermore, develop an active curiosity. Read for pleasure and inquisitiveness habitually. Browse your library, including math and science books, and the applications thereof. Explore new activities. Work different jobs. Try a new sport, game, other extracurricular activity, make new friends, or attempt a novel way of doing anything. Leave your comfort zone. There's a risk to everything, but you'll be surprised at the results from taking calculated risks, not foolish ones.

> *Employ your time in improving yourself by other men's writings, so that you shall gain easily what others have labored hard for.*
> - Socrates

By continuing to study correctly in all subjects, developing an active curiosity, becoming a habitual reader outside of the classroom on your own volition, staying active in sports and/ or other extracurricular activities, hanging around with good friends, and working a part-time job, your own wisdom and judgment will increase with time, and you will improve.

We hear a lot about *mindfulness* these days. If you want your study habits to improve and have a true desire to learn, find that study location covered earlier, then shut off all electronic devices while reading and doing your homework. You will start to experience the pleasure of silence, no interruptions, and being *mindful*. For those of you who say, "*I work better with music*", or "*I need to have the TV on while I study*", or "*I can text my friend and still get my work done*", these assertions are highly contestable, regardless of your test scores or grades. Like the athlete who is naturally strong or talented but doesn't work that hard yet is still a star player, he would be even that much better if he worked harder, was more focused, and adopted the correct attitude and approach.

Nonlinear Learning. Forgive me for this repetition, but the learning process in any subject is not always so straightforward. Learning, as with life's twists and turns, is many times irregular and *nonlinear*. School is the closest that you will come to a linear, organized, methodical, time-allocated learning process by design. To the extent that we can keep it linear to achieve our goals, we do that by persistence and perseverance, and taking advantage of what a formal education in controlled settings offers. You won't have the luxury of a formal education forever, at least not without cost. The opposite to formal schooling is experience brought up earlier. Consider all forms of learning as pieces of a puzzle. This is how one learns math, or any subject for that matter. It's like detective work. Each learning piece—formal vs. experience—has its pros and cons. Where one is deficient, the other can fill in the gaps. But both are needed to complete the puzzle as it were.

Be prepared to be disappointed or to have a degree of confidence in math that will later be squashed. Expect frustration, scrambled thoughts, and fatigue, especially late at night. Working hard, even if immediately unfruitful, produces desired results later. Sleep and sticking to a well-planned schedule that includes ample time for recreation works wonders. This is all part of the learning process. This same nonlinear learning process is needed for thinking professions in the work world as well. In fact, it's the predominant form of learning. Keep these words of wisdom in mind from Lucretius[7] and Poor Richard, respectively, during your times of studying frustration:

- *"The fall of dripping water hollows the stone"*
- *"By diligence and patience the mouse ate in two the cable; and Little strokes fell great oaks"*

Related to math study and learning methods is the subject of how to teach math properly. This has been a subject of debate for years in education circles. (It continues to be, and probably always will be.) You have my condolences once again if you are being used as a guinea pig in any new teaching method being implemented by a school system or maverick teacher. I've been there. I recall some early Charlie Brown cartoons during my childhood summer reading in

---

7   Titus Lucretius Carus, Roman poet and philosopher, 1st century BCE

the 1960s where Charlie Brown's teacher was introducing *new math*, which only confused poor Charlie Brown and his friends. "New" teaching methods do not negate this chapter's lessons. Don't allow yourself to be duped. In another story from fifth grade (1971-1972), my school was introducing the latest fad of *speed reading* and its techniques. Reading skills can be improved, and your reading speed can be increased. But I knew then, and I think my other classmates knew at our young age, as well as the teacher who was being forced to teach it, that the particular method was ineffective nonsense.

> *Genius is 1 percent inspiration and 99 percent perspiration.*
> - Thomas Edison, American inventor, 1847-1931

Chapter Wrap-up

When considering how to study and learn math, learning how *not* to study and learn math can help. I—in junior high school and high school—was the poster child for how *not* to approach academics: disorganized; never reading the book; not paying attention or taking notes as much as I should have; doing my homework only some of the time; not preparing for exams or doing so improperly; procrastinating constantly; watching too much TV; having too much fun; and approaching math and academics with the wrong attitude. This is no way to succeed. You don't need to join the Marines to "get religion." Turning your mathematics life around is not too late, but the sooner the better. Merely resolving to work harder is not the solution. (I attempted it multiple times then.) Increasing your efficiency and effectiveness by following the six basic steps covered in this chapter is the way to go.

Regarding the sixth step, *Change your attitude and approach*, if you don't agree with it, therein lies the problem. Don't agree at your peril. Make things up as you go along and continue to be at a disadvantage. (This applies to life, in general, as well.) As has been said over the years and as I said before, *Attitude is everything*. Three very important words. They apply everywhere. Now is the time at your young age to realize the power of these words, adopt them, and to start approaching life from an advantage. Some people with long lives never learn this, and they suffer the consequences.

6

Final Words

What we know is not much. What we do not know is immense.
- Pierre-Simon Laplace, French mathematician, astronomer, and physicist, 1749-1827

Formal definitions of mathematics from experts have been presented, and I added my supplementary definition (shortened here) in that it is also a language and set of tools. We reviewed that math is used to allow or help us solve real-world problems and make things work. Even if we don't use math actively and directly, we certainly all use it or benefit by it passively or indirectly. The three main reasons for needing math are national security, individual survival, and to avoid being taken advantage of.

In the final chapter, we covered how to study and learn math using these six components:

1. Get organized and manage your time
2. Read the book
3. Pay attention, take notes, and ask questions
4. Do the homework
5. Prepare for exams properly
6. Change your attitude and approach

These six studying and learning components can be summarized with this principle: Get disciplined, stay engaged, and do *all* the work—assigned and unassigned. In the short term, expect to be entertained and expect to be unfulfilled. In the long term, you will gain insight, relative peace of mind and self-confidence, valuable knowledge, and ultimately academic success.

Harping. Math is not a silver bullet. It will not make you healthy, wealthy, and wise by itself. But it is a necessary ingredient to help you get to where you want to go in life. It's part of a balanced education, and learning it is not always enjoyable.

Math requires prerequisites. You can't jump into any old math course, or a course in science, finance, and the like without an adequate math background. A scientific calculator or computing software can't be used to its full extent without sufficient math; most functions will be useless to the uneducated user. The manipulation of fractions, graphs, and positive and negative integers that you learned about a few years ago, still matter and always will. All topics covered in previous math courses must be mastered to move forward. The more you progress in math (and its dependent subjects), the more you will come to appreciate it. You'll develop in-depth skills that will help you reach your potential in sometimes subtle and unobvious ways.

Experience Matters. A formal education is critically important to your future, but it is not the only factor. Recall from Calvin Coolidge's quip on persistence: "*...the world is full of educated derelicts.*" Excluding work ethic and other character traits here, experience matters and can be attained from various jobs, sports, other extracurricular activities, playful experimentation, emotionally painful and joyful events, travel, meeting new people, friends and enemies, reading, doing *math homework*, and so on.

> *You don't know what you don't know.*
> - Common expression

We hear a lot about STEM courses in school, but what's missing in those conversations is the value derived from directly related industrial arts and vocational training that applies to practical problem-solving. The good news for current and future engineering students is that many colleges and universities have acknowledged past deficiencies in their curricula and are requiring more hands-on work in labs and machine shops, as opposed to mostly analytical classroom-type learning.

Other often overlooked parts of experience are the value derived from work drudgery and dealing with the public. Many professions entail work filled with daily, painful grind that comes in a variety of forms. That's why it's called work, and why people don't take vacations to go to work. Remember that there's virtue in drudgery, much can be gained from it, and life is full of it. What's more, by working with the public, one can gain effective people skills, and the realization of how bad people can be (the few)—and how good (the most).

Math skills matter as do others. Employers are constantly on the lookout for hardworking and energetic people with a proven track record who will add value to their organizations and not simply be "wage earners". In business, what matters most is the bottom line, namely, profits, for there can be no business without profits. Good employees can add to them. Every industry that features complex products requires an incredible amount of specialized knowledge that can only be acquired by working in that industry for an extended period of time. It cannot be learned in a classroom. We want to avoid succumbing to the fallacy of overconfidence in academic credentials and/or being overly impressed by them. One of my UMass Amherst professors brought this point home when he told our class of mechanical engineering seniors who were on the verge of graduating about how much knowledge we had versus the knowledge needed in the work world: *"You seniors don't know s—t."* You will find this to be very true when you enter into industry. Prepare to be humbled, even if you've earned your doctoral degree. I recall a FedEx TV commercial some years back when an experienced woman in a particular company was assigning a task to a new hire, where the new hire thought the task was beneath him due to his academic achievement. Before starting the task, he stated indignantly, *"I have an MBA"*, which is when the woman responded, *"In that case, I'll have to show you how to do it."* I thought it was pretty funny. Most people could care less about your MBA or doctoral degree. The good news is that learning becomes easier as you mature in the work world. You will get "smarter" in how to learn, which becomes very important when switching jobs and/or employers.

The Smart Kids in Class. Learn not to be intimidated by the "smart" students in your math class or being in fear of sounding stupid. Sometimes the smartest math student is not the smartest person. The really smart student is the one who understands the importance and relevance of math, and its necessity to the physical sciences, life sciences, social sciences, engineering, finance, trades, and just about everything else in life. The smartest student is also the one who can *apply* math to the real world. It's the smart student who is not falsely proud of how much knowledge he or she has accumulated but is humbled by how much he or she doesn't know. Also recognize that you may have an outstanding attitude, exceptional people or "soft" skills, and a stronger work ethic than the so-called math whizzes. These qualities will serve you very well for the rest of your life in all aspects. Conversely, acknowledge that sometimes the smartest math students are, in fact, highly intelligent, hardworking, likable, tolerant of the not-so-math-proficient students, and well-balanced individuals with superior attitudes. It's these students whom you can be impressed by and want to emulate.

Parting Remarks. I hope that you learned something from this book, and can apply its wisdom not only to math but to the work world and other facets of life. If my ultimate purpose of this book has been fulfilled, you may not love math, but you don't hate it either. Additionally, you understand what it is, what to do with it, the reasons for studying it, and the best ways to do so. As a result, you can now make more informed choices about future courses of study, college, and career paths. Ideally, you have a new lease on life.

As someone once said, "*Youth is wasted on the young*" (often attributed to Irish playwright George Bernard Shaw, 1856-1950). Use the example of my academic youth for what *not* to do. Learn from my all-too-common mistakes, and those of others—inside and outside of school. You'll learn a lot. Be wary of the "cool" kids, especially those who spell trouble.

Continue to learn throughout life. Math and school, in general, are not wastes of time. They're far from it, but they're not everything. Experience, which is to say, an informal education and OJT, plays a critical role in learning. You need "all the above" as you progress. Both a formal education and an informal one will help open your eyes to everything else. There's so much to learn in life, and so little time. For those occasions when you're not in school or working, please don't use them as a license to stop learning and do nothing. From Poor Richard, "*A life of leisure and a life of laziness are two things*" and "*Industry need not wish.*"

If you remember nothing else from this book, remember this and burn it in your brain—for math and everything else in life:

Attitude Is Everything

Appendix

Ten things to tell yourself to help ensure that you will be unsuccessful in math:

1. *I don't care about what math is or where it comes from.*

2. *Other than the basics, no one really uses this stuff. It's just a bunch of fancy symbols and numbers. I'll never use it again.*

3. *There's no point in continuing math beyond what is required to graduate.*

4. *I'm wise to the world and how it works. No one is going to take advantage of me because I didn't have enough math skills or take math seriously.*

5. *I already study for exams. I review my notes, check over a few things, and that's about it. There's not much more I can do. By this point, you either know it or you don't.*

6. *I'm already organized with an overloaded schedule. I just don't have any more time to go above and beyond assigned work. Hmmm...My friend is texting me, I have to check my social media account, and my favorite show is on TV in about ten minutes.*

7. *I don't need to read the textbook. I just skim the relevant content and do the assigned problems. This "system" has always worked for me in the past...kind of.*

8. *Sitting in front, paying attention, taking notes, asking questions...These things are for nerds. I don't want to stick out or have anyone think that I'm a geek or that I care about math.*

9. *Some students, the really smart ones, get this stuff right away. I don't. I have to work much harder than they do to understand it, which tells me that they're smart, I'm dumb, and my efforts are futile.*

10. *My friends and family don't have a negative influence on me and my approach to math and schoolwork, in general. And even if they did, there's not much that I can do about it.*

Index